T0321169

Magnetic Phase Transitions in Single Crystals

Magnetic Phase Transitions in Single Crystals

David P. Belanger

University of California, Santa Cruz, USA

NEW JERSEY · LONDON · SINGAPORE · BEIJING · SHANGHAI · HONG KONG · TAIPEI · CHENNAI · TOKYO

Published by

World Scientific Publishing Co. Pte. Ltd.

5 Toh Tuck Link, Singapore 596224

USA office: 27 Warren Street, Suite 401-402, Hackensack, NJ 07601

UK office: 57 Shelton Street, Covent Garden, London WC2H 9HE

British Library Cataloguing-in-Publication Data
A catalogue record for this book is available from the British Library.

MAGNETIC PHASE TRANSITIONS IN SINGLE CRYSTALS

ISBN 978-981-125-948-7 (hardcover)
ISBN 978-981-125-949-4 (ebook for institutions)
ISBN 978-981-125-950-0 (ebook for individuals)

For any available supplementary material, please visit
https://www.worldscientific.com/worldscibooks/10.1142/12923#t=suppl

Typeset by Stallion Press
Email: enquiries@stallionpress.com

Printed in Singapore

Preface

In the past few decades the phenomena associated with phase transitions have been investigated in experiments, theoretical studies, and computer simulations. Much of the research involves magnetic systems because the universal properties of model Hamiltonians are surprisingly often realized in high quality magnetic crystals. A book that comprehensively reviews every aspect of phase transitions in magnetic crystals would be a daunting task and that is not the purpose here. Instead, this book is a more narrow review of experimental work in several major areas that serve as models for classes of critical behavior and related phenomena encountered in magnetic crystals. Well-characterized magnetic systems are covered for each area, allowing a deeper and more accurate understanding of the experiments and their impact. The book will not cover the vast subject of magnetic films, but the understanding of phase transitions in bulk crystals can contribute greatly to the understanding of the more complicated magnetism encountered in thin film research. Most of the systems covered here are insulators because the interactions are short-ranged, more readily allowing the asymptotic critical behaviors to be accessed.

In Chapter 2, formal definitions of the basic critical behaviors will be given that are essential to extracting the universal critical parameters from experimental data. Some details will be postponed until the relevant systems are discussed.

In Chapter 3, useful background will be given for some techniques that are used to characterize phase transitions in antiferromagnets. The amount of detail will vary because some techniques need less explanatory information while others need details that justify the interpretations of the data collected using them. Special attention is given to thermometry and specific heat techniques because some experiments can be more accurately

v

interpreted if they are more fully understood. Other areas, such as neutron scattering have been well reviewed previously.

Chapter 4 covers experiments in isotropic and anisotropic pure $d = 2$ and $d = 3$ Ising antiferromagnets. For $d = 3$, the effects of crossover from Heisenberg to Ising behavior are analyzed. A large amount of detail is given for these systems because they are well characterized and the results serve as a basis for understanding more complicated phase transitions. The effects of magnetic dilution on the universal critical behavior will be examined. By applying a uniform field to these same systems, the random-field Ising universality class can be studied. This area of study is ongoing, but the results up to the present will be described. Finally, the characterization of the critical behavior of isotropic magnets will be covered.

Chapter 5 will cover experiments exhibiting behavior associated with the magnetic percolation threshold concentration. This includes energy excitations, both Ising in strongly anisotropic magnets and fracton excitations in isotropic system. Phase diagrams that exhibit spin-glass-like-like behavior will be discussed. Low temperature domain dynamics in anisotropic magnetic systems will be reviewed.

Chapter 6 will cover the effects of strong frustration on uniform antiferromagnetic systems. The frustration originating from the crystal geometry or conflicting interactions can lead to new universality classes.

Chapter 7 is a short chapter describing a possible example of surface ordering that leads to new universal critical behavior that is normally difficult to detect. It is an example of how characterizing critical behavior can help to identify the source and nature of magnetic order.

Chapter 8 will summarize some general ideas that were developed in the earlier chapters and pose some open questions that will hopefully be addressed by future experiments, theory and computer simulations.

I have been fortunate to have had the opportunity to collaborate with talented researchers on experimental investigations of magnetic order in crystals. I have also enjoyed working with extraordinary theorists who have modeled the systems and provided valuable insights into the associated phenomena. I hope this book will serve in some way those who continue research on the fascinating topics associated with phase transitions and order in magnetic crystals. I also hope that it will provide an introduction to researchers interested in entering studies of phase transitions in magnetic crystals.

David P. Belanger

Acknowledgment

I would like to thank the American Physical Society, Elsevier, IOP Publishing, the Physical Society of Japan, Springer, Springer Nature, the American Institute of Physics and EDP Sciences for permission to reproduce figures.

Contents

Preface   v

Acknowledgment   vii

1. An Introduction to Phase Transitions and Universality   **1**

**2. Universal Critical Behavior from Theory
   and Simulations**   **15**

3. Background on Experimental Techniques   **23**

   3.1 Thermometry   24

   3.2 Specific Heat Techniques   26

   3.3 Optical Measurements of Critical Behavior   32

       3.3.1 Optical birefringence   33

       3.3.2 Specific heat critical behavior via birefringence
              techniques   35

       3.3.3 Faraday rotation   41

       3.3.4 Isotropic magnets and the birefringence
              technique   43

   3.4 Capacitance   43

   3.5 Susceptibility and Thermal Expansion   44

   3.6 Neutron and X-ray Scattering Measurements   46

       3.6.1 Elastic scattering   47

       3.6.2 Inelastic scattering   48

   3.7 The Effect of Concentration Gradients on Critical Behavior
       Characterizations   48

4. Critical Behavior Experiments on Anisotropic and Isotropic Antiferromagnets   **57**

4.1   Crystals   58
     4.1.1 $d = 3$ Anisotropic crystals   58
     4.1.2 $d = 2$ Anisotropic crystals   61
     4.1.3 $d = 3$ Isotropic crystals   61
4.2   Phase Diagram Measurements of Anisotropic and Isotropic
     Antiferromagnets   62
     4.2.1   The bicritical and tricritical points in pure and
             dilute anisotropic antiferromagnets   62
     4.2.2   The influence of anisotropy on the
             transition T_N   71
4.3   Specific Heat Critical Behavior of Pure Ising
     Antiferromagnets   76
     4.3.1 $d = 2$ Pure Ising specific heat critical
             behavior   76
     4.3.2 $d = 3$ Pure Ising specific heat critical
             behavior   77
4.4   The Order Parameter of Pure Ising Antiferromagnets . . .   83
     4.4.1 $d = 2$ Pure Ising order parameter   83
     4.4.2 $d = 3$ Pure Ising order parameter   84
4.5   Neutron Scattering Critical Line Shapes of Pure Ising
     Antiferromagnets   86
     4.5.1 $d = 2$ Pure Ising neutron scattering   86
     4.5.2 $d = 3$ Pure Ising neutron scattering   91
4.6   The Random-Exchange Ising Model   98
     4.6.1 $d = 2$ Random-exchange Ising model   99
     4.6.2 $d = 3$ Random-exchange Ising order parameter . . 104
     4.6.3 $d = 3$ Random-exchange Ising specific heat critical
             behavior 105
     4.6.4 $d = 3$ Random-exchange neutron scattering 106
     4.6.5 $d = 3$ Random-exchange dynamics 108
     4.6.6 $d = 3$ Random-exchange susceptibility 109
4.7   The Random-Field Ising Model 110
     4.7.1 $d = 2$ Random-field Ising critical behavior 115
     4.7.2 $d = 3$ Random-field Ising critical behavior 118
     4.7.3 $d = 3$ Random-field dynamics for $x < x_v$ 126

      4.7.4   Vacancy percolation and the stability of $d = 3$
               antiferromagnetic long-range order 132
      4.7.5 $d = 3$ Random-field order parameter 136
      4.7.6 $d = 3$ Random-field specific heat 143
      4.7.7 $d = 3$ Random-field neutron scattering
               line shapes 145
  4.8 $d = 3$ XY Specific Heat Critical Behavior 155
  4.9   Experiments on Isotropic Magnets 156
      4.9.1   Optical birefringence in isotropic
               antiferromagnets 156
      4.9.2   The specific heat critical behavior of
               isotropic magnets 159

5. **Domains, Excitations, and Spin-Glass-Like Behaviors**   **161**

  5.1   Domain Structure Dynamics at Low Temperature 161
  5.2   The Phase Diagram of Anisotropic Antiferromagnets
       Above the Magnetic Percolation Threshold
       Concentration 165
  5.3   Excitations Near the Magnetic Percolation Threshold
       Concentration 173
      5.3.1   Fracton excitations in isotropic diluted $d = 2$
               and $d = 3$ antiferromagnets near the magnetic
               percolation threshold concentration 174
      5.3.2   Excitations in anisotropic $d = 2$ and $d = 3$
               diluted antiferromagnets near the magnetic
               percolation threshold concentration 176

6. **Experiments on Pure Magnets with Frustration**   **179**

  6.1   The XY Stacked Triangular Lattice 180
  6.2   Examples of Other Chiral Systems 184
      6.2.1   Holmium 184
      6.2.2 VF_2 186
  6.3   Ising Stacked Triangular Lattice 186

7. **The Unusual Magnetism of $LaCoO_3$:**
    A Thermally Excited Exchange Interaction
    and Ordering at Twin Interfaces   **193**

8. Conclusions and Outstanding Questions   199

8.1   Overall Summary of the Results of the Experiments . . . 199

8.2   Some Open Questions About Equilibrium in the
      Random-Field Ising Model 202

8.3   Future Work 205

Bibliography   207

Index   227

Chapter 1

An Introduction to Phase Transitions and Universality

Pure, unfrustrated magnetic systems have been comprehensively characterized and they serve as models for exploring phase transitions in more complicated magnetic systems that have intrinsic frustration and those with quenched disorder. The pure crystals without frustration are high in quality and critical behavior data can be analyzed exceptionally close to the transition temperature. As a result, the agreement between experiment and theory is extraordinarily good. These same systems can be magnetically diluted to serve as models of the effects of quenched randomness on static and dynamic critical behavior, the dynamics of domain walls, and percolation threshold phenomena. Systems that are not diluted but have significantly frustrated interactions exhibit new classes of critical behavior. Each of these areas will be covered in this book.

Because there are many reviews of phase transitions from the perspective of theorists, this book is written from the perspective of an experimentalist. Hopefully, that will serve three purposes. First, the inclusion of detail about some of the experimental techniques can serve as a guide for experimentalists studying critical behavior close to magnetic phase transitions. Second, it might help theorists better understand and evaluate reports of experimental results. Finally, it might highlight some areas of research where the interpretations of experimental data rely on support from theory and simulations for the best analyses of data. In some cases, work is yet to be done to better integrate experimental and theoretical interpretations.

Figures 1.1 and 1.2 show generic features of a gas/liquid phase diagram and of a phase diagram of an Ising ferromagnet, respectively. A large part

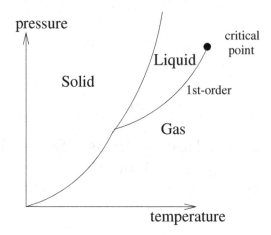

Fig. 1.1: Features of a pressure–temperature phase diagram. The first-order transition boundary between the liquid and gas phases ends at a second-order transition critical point. Crossing the first-order transition boundary involves a latent heat and coexistence of the two phases at the boundary. The behavior crossing the critical point can be described in terms of power laws as discussed in Chapter 2. The difference between the gas and liquid phases becomes smaller across the first order boundary as the critical point is approached and the transition becomes continuous crossing at the critical point. At high enough pressures and temperatures, the change from gas to liquid can take place continuously without going through a transition. The point where the three phases meet is a tricritical point.

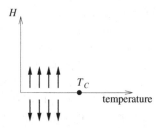

Fig. 1.2: The H versus T phase diagram of a ferromagnet. The critical point in this case is at $H = 0$ and $T = T_C$. The first-order line below T_C separates the ordered phase that has net spins up from the one with net spins down. The critical point cannot be crossed with $H \neq 0$.

of this book focuses on the behavior near the critical point; a second-order phase transition takes place as it is crossed. The $d = 3$ Ising transition and the $d = 3$ liquid–gas transition share many universal features that do not depend on the details of the systems. It is such universal properties that allow phase transitions that are observed in experiments to be characterized and compared with theory and simulations through the use of idealized models. Details of the experimental systems and the simulations do not need to be identical. The comparisons can either verify the understanding of the universal properties of the phase transitions or they can highlight areas where further work needs to be done to arrive at a full understanding.

Magnetic materials play a significant role in technology and are important to understand for that reason. However, their importance in the context of this book is that they provide impressive opportunities to probe the ordering process that takes place near magnetic phase transitions and to compare the observed behaviors with theoretical models. The results are applicable to phase transitions in a broader sense because of universality. Magnetic materials that are fascinating but do not exhibit transitions and are not related to the materials that do will not be covered. An example of what will not be covered is the extraordinary phenomena associated with canonical spin-glasses where no magnetic transition takes place. The book will concentrate on magnetic crystals, excluding, for the most part, other interesting magnetic systems that do exhibit phase transitions such as thin films and nanoparticles. Those systems tend to be more complicated and difficult to interpret without a basic understanding of the transitions that take place in the corresponding large crystals that can be more readily modeled.

While transitions occur in a wide variety of physical systems, magnetic crystals provide many unique opportunities that are difficult to realize in other systems such as liquids. The $d = 3$ Ising magnets have the same critical properties close to the phase transition as some liquids, but magnetic crystals present an array of different symmetries and properties that are difficult to realize and explore in other systems. For example, quenched impurities can be studied in magnetic crystals that are grown from chemical mixtures with magnetic and nonmagnetic constituents, something that cannot be readily accomplished in liquids.

A variety of experimental techniques can be used to characterize the symmetries and other properties of crystals. This allows the accurate modeling of transitions that take place in them. As a phase transition is approached, the behavior becomes simple in the sense that the behavior is governed only by the basic symmetries such as whether the geometry of the interactions in the system are best described as two-dimensional ($d = 2$) or three-dimensional ($d = 3$). Another symmetry that defines the class of behavior near the transition is defined by whether the spins can orient in all directions (Heisenberg), preferentially along one direction (Ising), or only within a plane (XY). Other basic properties such as the kind of disorder, for example, random-exchange or random-field, can also determine the universal properties of the critical behavior near the ordering transition. Frustrated interactions can also greatly influence the fundamental nature of the magnetic transitions.

Other than such basic symmetries and properties, the universal parameters of a phase transition do not depend on details of the system as long as the transition is approached closely enough. The length scale for thermal fluctuations in the system, ξ, must be larger than all relevant interactions for the behavior to be representative of the universality class determined by the basic symmetries. One great advantage to studying phase transitions in insulating magnetic crystals is that the interactions between moments are often limited to closely neighboring ions. The strength of the interactions typically fall off dramatically with the distance between the moments because it is determined by overlap of adjacent ion orbitals. This property of magnetic systems, for example, is what allows the study of ordering in systems that are effectively two-dimensional when planes of strongly interacting moments are fairly well separated from other planes.

The short-range nature of interactions in an insulating magnet often allows measurements close to the transition to reflect the asymptotic critical behaviors predicted by theories and computer simulations. However, even with insulating materials, there can be strong crossover effects as the transition is approached and this will be discussed in the case of the weakly anisotropic antiferromagnet MnF_2, which behaves as an Ising system only extremely close to the transition. It will be compared to the strongly anisotropic FeF_2 system that exhibits Ising behavior much further from the transition temperature. Crossover effects are also manifest in systems with magnetic dilution, such as the random-exchange and random-field Ising model systems. The new critical behavior for the $d = 3$ random-field Ising transition is only observed close to the new transition temperature and that often requires excellent thermometry and crystals with a high degree of chemical homogeneity.

The universality of the critical behavior close to a transition also allows computer simulations to be done with the simplest Hamiltonian that reflects the important symmetries. It is the goal of theory, simulations, and experiments to determine the universal parameters determined by the overall symmetries. When these different approaches come to a consensus, it usually means the transition is well understood. In the absence of consensus, it signifies that there are still things to study and model. Antiferromagnetic crystals have an impressively wide variety of symmetries and interactions. Because of that, a surprising number of universality classes can be realized and studied in antiferromagnetic crystals.

A familiar magnetic material commonly experienced in everyday life is the ferromagnet iron. Pure iron is strongly ferromagnetic at room temperature; it disorders on a local scale at temperatures $T > T_C = 1043$ K. However, at room temperature, it does not attract other pieces of iron as we might expect from the local ordering. This is a consequence of domain formation. The magnetic field of a ferromagnetically aligned region extends far outside of that region and that costs an amount of energy proportional to the square of the field strength integrated over space. To avoid that energy cost, iron forms local domain structures with internal ferromagnetic ordering in directions that vary from domain to domain. The field outside the iron crystal cancels as the moments from all the domains average to zero, thereby lowering the energy. However, the formation of domains also has an energy cost because the moments at the domain wall interfaces are not fully aligned. The domain size is determined by the balance of domain wall energy and the energy cost of extended magnetic fields. As long as there is no external field applied, the iron crystal will have no net macroscopic moment.

When an external field is applied to pure iron, the domain walls shift so that domains ordered with moments favorably aligned with the field grow and domains with unfavorably aligned moments shrink. The iron crystal then behaves macroscopically as a magnetic crystal. Once the external field is removed, the domain walls revert back to their original configuration and the iron is macroscopically nonmagnetic once again.

The situation changes if there are impurities or defects in the iron. This is the situation in steel, for example, where the impurities strengthen the iron. Most forms of steel can be magnetized in an externally applied field, but when the field is removed, some of the macroscopic magnetism remains. The retained magnetism is a result of the energy barriers the impurities create that pin the domain walls. The macroscopic moment can be decreased by mechanical shock or eliminated by heating it to temperatures above T_C. Above T_C, the local moments are disordered. As the crystal temperature is decreased through the transition, the moments in each domain must choose a direction in which to order. That locally breaks the spatial symmetry below the transition. It is the symmetry-breaking phase transition we are most interested in for most of the systems covered in this book, but domain dynamics in the ordered state are also of interest.

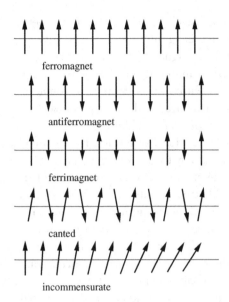

ferromagnet

antiferromagnet

ferrimagnet

canted

incommensurate

Fig. 1.3: A few simple one-dimensional depictions of possible spin orderings. The ferromagnet has all spins ordering in the same direction. The antiferromagnet has spins ordering in alternating directions and is described as having two ordered sublattices. The ferrimagnet also has two sublattices, but the moments on the two sublattices are unequal, giving a net moment. The canted antiferromagnet has two noncolinear sublattices with equal moments on each, but the noncolinear order produces a net moment. The incommensurate case is when the change in direction of the moments varies with a wavelength that is independent of the underlying lattice periodicity.

Several kinds of order are useful to consider in the discussions in the following chapters. The systems that undergo transitions and can be studied are two- and three-dimensional, but Fig. 1.3 depicts some of them in one-dimensional representations. Some of these, the ferromagnet, ferrimagnet, and canted antiferromagnet have net moments, even though the ferrimagnet and canted antiferromagnet have interactions that are primarily antiferromagnetic. The antiferromagnet and incommensurate magnet have no net moment. Ordering in real crystals can be much more complicated than the depictions. Ferromagnets typically have second-order transitions, but only for zero applied field. Antiferromagnets do have transitions in a uniform applied field and can show bicritical behavior, as will be discussed. One type of order that will be discussed, but is not in the figure, is that of a metamagnet. $FeCl_2$ is an example where well-ordered ferromagnetic planes interact antiferromagnetically and the phase diagram shows a tricritical point.

Although ferromagnets are technologically important, antiferromagnets are better suited for the study of universal critical behavior near transitions. Ferromagnetism is far less common than antiferromagnets in insulating materials. Although domains form in antiferromagnets, it is not to eliminate magnetic fields outside the crystal, as is the case for ferromagnets. It is also difficult to realize many model phase transitions in ferromagnets. For example, magnetic fields provide an excellent probe in antiferromagnets, but the phase transition critical point is not even crossed for a ferromagnet in a field because the symmetry has already been broken by the applied field above T_C. Symmetry breaking in a ferromagnet only happens when cooling the crystal with $H = 0$. Antiferromagnets, on the other hand, do undergo phase transitions in an applied field, thereby opening up the study of many features of phase diagrams inaccessible to ferromagnets. For that reason, the systems featured in this book are antiferromagnetic.

Critical behavior can be measured much closer to the transition temperature in liquid helium than in magnetic crystals and in some cases such studies are superior to experiments on the same universality class in crystals. However, many more universality classes can be realized in magnetic systems. Magnetic crystals have a distinct advantage in studies involving quenched randomness. Examples that we will examine include random substitution of magnetic ions, such as Fe^{2+} with diamagnetic ions such as Zn^{2+}. Crystals can be grown from chemical mixtures such as FeF_2 and ZnF_2 with good local homogeneity although, in practice, the macroscopic uniformity is often the limiting factor in probing critical behavior close to the transition temperature. Great effort has gone into improving chemical homogeneity, but careful consideration of inevitable concentration gradients must guide the interpretation of critical behavior data. Some of the issues of concentration gradients when studying the random-exchange and random-field Ising models will be discussed in detail.

Magnetic crystals can be studied with an array of experimental techniques including neutron and X-ray scattering, specific heat, optical techniques, magnetometry, dilatometry, electron and nuclear magnetic resonance (NMR), and Mössbauer spectroscopy. Almost any sharp critical behavior observed in experiments can be related to just a few universal behaviors, including the specific heat critical behavior, the susceptibility, the order parameter critical behavior, the critical divergence of the fluctuation correlation length, and the dynamics of the transition. From these experimental techniques, the magnetic systems can be characterized with great detail. The critical behaviors near the transition can then be compared to the theoretical models for the behaviors of the appropriate universality class. As the comparisons are made, agreement can often be found between theory and experiment, giving confidence in our understanding of the transition being investigated. Sometimes, for example in the early days of studies of the random-field Ising model, profound discrepancies showed up between interpretations of experiments and the predicted behavior. In this case, the discrepancies led to a deeper understanding of the random-field phase transition.

Phase transitions cannot take place in one-dimensional systems or in two-dimensional Heisenberg magnets; thermal fluctuations are too great for long-range order to be established. This book will therefore not address such systems or any other system that is not at least closely related to possible magnetic phase transitions. Magnetic and nonmagnetic properties of some crystals will be described when that leads to a better understanding of the phase transitions in the crystals.

A comprehensive review of many theoretical and numerical aspects of phase transitions and critical behavior is given in Pelissetto and Vicari (2002), a review that includes extensive references to experiments done with zero applied magnetic field, including frustrated magnets and randomly dilute magnets. It does not address random-field systems. There are, however, many works, including reviews, of numerical and theoretical work on the random-field Ising system. We will reference these works and reviews when useful.

Figures 1.4–1.6 illustrate some basic features of systems near phase transitions using a simple example of a two-dimensional Ising ferromagnet. The images are generated using a Monte Carlo simulation of a 128×128 lattice with nearest-neighbor interactions, zero applied field, and periodic boundary conditions in which the spins on the top row interact with those on the bottom row, and the spins on the right edge interact with those on the left edge. Each site is accessed randomly and a Metropolis algorithm is used to attain thermal equilibrium. The three figures show the behavior at three different temperatures. The black squares represent "down" spins and the white squares are "up" spins. Each spin interacts with and tries to align itself with its nearest neighbors. The algorithm introduces randomness that properly simulates the thermal energy and tries to disorder the spins in a way that properly reflects thermal equilibrium. The simulation runs until an equilibrium configuration is reached. With modern laptops and desk computers, this does not take long. Figure 1.4 illustrates the fluctuations

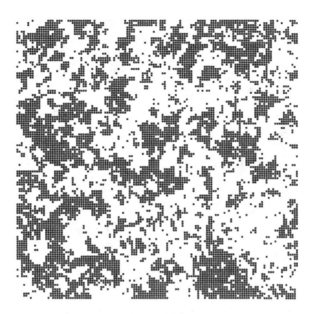

Fig. 1.4: Monte Carlo simulation (Barber, 1999) of a $d = 2$ Ising ferromagnetic lattice with periodic boundary conditions in zero field. White represents "up" spins and black represents "down" spins. The temperature is above the transition temperature. The reduced temperature is $t = (T - T_C)/T_C = 0.1$, where T_C is the transition temperature. At temperatures $T \gg T_C$, the thermal fluctuations dominate over the interactions between the spins and they cannot align with their neighbors. At the temperature depicted in this figure, there are local correlations among the spins as they attempt to order. However, the temperature is too large for ordering across the sample and there is a characteristic largest length scale for the fluctuations, ξ. The figure represents one moment in time. The fluctuations evolve on a time scale that grows with the size of the correlated regions. The average moment fluctuates around zero. As the system size grows, the average moment approaches zero. There is no net moment and therefore no preference for the orientation of the spins; the symmetry between up and down is not broken at this temperature. As t decreases, the size of the largest correlations grows, but the time average of the fluctuating moment remains zero.

at a temperature T just above the transition T_C with a reduced temperature $t = (T - T_C)/T_C = 0.1$. The net moment is close to zero because the number of up and down spins is approximately equal. The configuration changes with time, but the net magnetization averages to zero with small fluctuations over time because the symmetry between up and down is not broken at this temperature. The spatial size of the fluctuations varies from one spin up to a characteristic length, the correlation length for fluctuations, ξ, that increases as T decreases and approaches T_C.

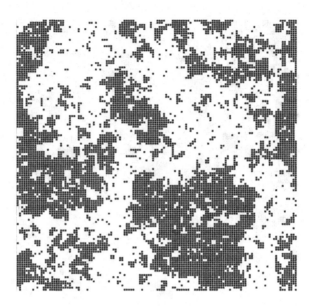

Fig. 1.5: The same $d = 2$ Ising simulation as in the previous figure, but at a lower temperature near the phase transition, with $|t| \approx 0$. The average net moment is still zero, but the size of the largest correlations, ξ, is as large as the sample. For an infinite sized sample, ξ becomes infinite, as does the time scale for fluctuations. When the temperature drops below T_C, the symmetry is broken and a net moment grows; the magnetization with no applied magnetic field will be either up or down, randomly.

The spatial fluctuations depicted in Fig. 1.5 are characteristic of $T \approx T_C$ ($t \approx 0$). The fluctuating regions of up and down spins vary with time with the larger regions evolving more slowly. The maximum fluctuation correlation length, ξ, becomes comparable to the sample size and would diverge in an infinitely large system. The characteristic time for fluctuations becomes very large for the largest fluctuating regions. In an infinite system, there is no characteristic length scale because ξ diverges and there are fluctuations on all length scales. The characteristic time scale for fluctuations also diverges as T approaches T_C in a system of infinite size. Although an infinite size cannot actually be achieved in real magnets, the size of magnetic systems in crystals is enormous compared to that achievable in computer simulations.

Pressure–temperature phase diagrams for a liquid–gas system typically have a first-order transition curve that ends at a point that is equivalent to T_C for a $d = 3$ ferromagnet and the universality classes are the same. In the case of the liquid–gas transition, a transparent liquid becomes cloudy

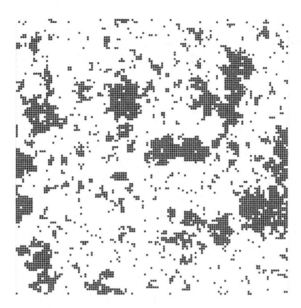

Fig. 1.6: The same simulation as in the previous two figures, but well below the transition with $|t| = -0.033$. A net moment has developed, with more up spins than down. The fluctuation length scale, ξ is limited and the fluctuations occur on a shorter time scale. As the temperature continues to decrease, ξ decreases and the net moment increases as the system becomes more ordered.

as the fluctuating regions of gas and liquid become large enough to scatter visible light (critical opalescence). In the same way, magnetic systems strongly scatter long wavelength X-rays and neutrons close to the transition and these techniques can be used to characterize some of the magnetic universal properties in the same way that light can be used to study universal behaviors near the liquid–gas transition.

Figure 1.6 shows the characteristic fluctuations below T_C, in this case with $t = -0.033$. The characteristic size of the fluctuations decreases as T decreases below T_C. At this temperature, there is a definite net moment, albeit with fluctuations, as up spins outnumber down spins. It is random which direction, up or down, has the majority of spins; the symmetry between up and down is broken only as T decreases through T_C. The net moment is the ferromagnetic order parameter. The fluctuations continue to decrease in size as T decreases. At $T = 0$, there are no fluctuations and all of the spins are aligned in one direction.

As the dimension of the system decreases, fluctuations play a greater role. In $d = 1$, thermal fluctuations are sufficiently strong that the Ising phase transition is completely suppressed and long-range order is never established; $d_l = 1$ is the lower critical dimension for uniform Ising systems. At large enough dimensions, fluctuations do not play a role in the critical behavior; $d_u = 4$ is the upper critical dimension for the uniform Ising model.

In an applied field, a net moment is induced in a ferromagnet at all temperatures and no symmetry breaking transition occurs. This is equivalent to the liquid–gas transition where one can change continuously from the gas to the liquid by doing an end-run around the critical point, avoiding the critical point and the first-order transition line. At temperatures $T < T_C$, the field can be reversed. As H passes through zero, the spins will flip direction, if they can equilibrate, in a first-order transition.

The transition in a $d = 2$ Ising system with ferromagnetic interactions in zero applied field belongs to the same universality class a $d = 2$ Ising system with antiferromagnetic interactions; the symmetries involved are exactly equivalent. In the case of the antiferromagnetic, neighboring spins align antiparallel. The order parameter is the staggered moment with the sign changing for alternating spins. The system can be viewed as having two sublattices that order in opposite directions. Unlike the case of a ferromagnet, magnetometry does not probe the order parameter, but the scattering intensity at antiferromagnetic Bragg point in reciprocal space does. Typically, antiferromagnetism doubles the lattice parameters of the unit cell, so Bragg scattering occurs at different locations in reciprocal space, typically at $1/2$ the spacing, than the atomic or nuclear Bragg points. A major difference between the ferromagnetic and antiferromagnetic transitions is that, unlike the ferromagnet, an applied field does not break the symmetry of the antiferromagnetic order above T_C; it is not conjugate to the order parameter. Note that it is usual to use T_N for the transition temperature in antiferromagnets in zero field, but it sometimes confuses the discussion and, for the antiferromagnet, T_C and T_N will be used interchangeably. That the antiferromagnetic transition can take place in an applied uniform field proves crucial in studies of several types of second-order transitions that will be covered in this book where the applied field is used to tune the properties of the phase transitions.

To fully characterize the transition, several universal properties need to be characterized as the phase transition is approached. For antiferromagnets, the order parameter is the staggered moment, M_S, and is not the net magnetization, M, of the entire crystal. It is zero for $T > T_C$ and grows as T decreases below the transition, typically as a power law in t. A variety of techniques can be used to characterize it, including neutron and X-ray scattering, NMR, Mössbauer scattering, and others that in some way couple to the antiferromagnetic order. Several of these techniques will be described for specific systems in this book. The characteristic correlation length for fluctuations, ξ, is typically determined from the line shape analysis of neutron scattering data. It diverges as T approaches T_C from above and below as a power law in t with the same exponent, but different amplitudes. For antiferromagnets, the susceptibility of the order to a staggered field, χ_S, will also behave as a power law in t, again with the same critical exponent, but different amplitudes above and below T_C. The staggered field comes from neutrons in scattering experiments and is related to the amplitude of the scattering line shapes. This is not accessible using magnetometry which detects the uniform susceptibility. The magnetic specific heat can be measured in antiferromagnets using various techniques such as pulsed heat, optical techniques, and magnetometry. The phonon background in pulsed heat specific heat measurements can sometimes overwhelm the magnetic contributions, making measurements of the magnetic critical behavior difficult and in such cases optical techniques might prove advantageous. The specific heat can exhibit a divergent power law in t, but it can also show a cusp behavior or a logarithmic divergence, depending on the universality class. Along with the static critical behaviors, the dynamic critical behavior can be determined using a variety of techniques. If all of the critical behaviors can be accurately measured using various techniques, the universality class can be well characterized. When the meaning is not ambiguous, the subscript indicating the staggered moment or susceptibility might be dropped for simplicity.

Chapter 2

Universal Critical Behavior from Theory and Simulations

Only a few basic symmetries and properties of a system determine the universality class of a phase transition. The properties close to the transition temperature can be completely characterized by a few critical parameters specific to the universality class that governs the transition, such as exponents and amplitude ratios of power law behaviors. The challenge for experimentalists is to measure the universal parameters and compare the values to those predicted by theory and determined from computer simulations. Beyond the properties that define the appropriate universality class, the details of the system are not important for the universal parameters. In principle, that makes the comparisons of experimental systems and idealized models straightforward. The details of the system, while not important to the asymptotic universal critical behavior, can make it more difficult to determine the universal parameters accurately in experiments. So, experimentalists need to be aware of them when attempting to characterize the universal parameters. Typically, the critical behavior must be measured close enough to the transition to ensure the fits to the data give accurate exponents and amplitudes of the asymptotic power law behaviors. In addition, accurate determination of the universal critical properties in experiments sometimes requires accurate models that can be used to analyze the data. An example of this is encountered in the analysis of neutron scattering line shapes. This topic will be discussed in some detail at appropriate points in the book to show the importance of having good models to analyze data and to show that not having good line shape models can limit the accuracy and scope of the data analyses.

The thermodynamics and statistical physics of phase transitions have been thoroughly reviewed in the literature, for example in Fisher (1967), Hohenemser *et al.* (1989), Collins (1989), Pelissetto and Vicari (2002), Hohenberg and Halperin (1977). The important quantities of magnetic systems that can be calculated and measured can be expressed in terms of the temperature T and field H. Because the critical behavior is universal close to the transition, examples of more appropriated variables might be $t = (T - T_C)/T_C$ and $h = (H - H_C)/H_C$. Using such dimensionless parameters removes the nonuniversal quantities T_C and H_C from the characterizations. In the equations below, the $+$ and $-$ symbols refer to $t > 0$ and $t < 0$, respectively.

The correlation length for fluctuations is given by

$$\xi = 1/\kappa = \xi_0^\pm |t|^{-\nu}, \tag{2.1}$$

and diverges as $|t| \to 0$ with a positive exponent ν. For ferromagnets, the order is the uniform magnetization, M, but the systems we will consider are antiferromagnets, in which case the order parameter is the staggered magnetization, M_S. The order grows below the transition as

$$M_S = M_0 |t|^\beta \quad \text{for } t < 0, \tag{2.2}$$

and is zero for $t > 0$. At the transition temperature of a ferromagnet, the magnetic moment grows with the field as

$$M \propto h^\delta. \tag{2.3}$$

For an antiferromagnet, the equivalent expression would relate M_S to an applied staggered field, which is not easy to do. For an antiferromagnet, the susceptibility of the order to a small conjugate field is the staggered susceptibility

$$\chi_S = \chi_0^\pm |t|^{-\gamma}. \tag{2.4}$$

The specific heat has the dependence

$$C = A^\pm |t|^{-\alpha} + B, \tag{2.5}$$

where α can be positive, zero, or negative. For the case of $\alpha = 0$,

$$C = A \ln(|t|). \tag{2.6}$$

For $\alpha > 0$, the specific heat diverges as $|t| \to 0$. If α is negative, the specific heat is a cusp with the value B at $|t| = 0$ and, in that case, the constant B

is an essential part of the critical behavior. In many cases, but not all, the critical dynamics can be expressed by the power law

$$\tau \propto \xi^z, \tag{2.7}$$

where τ is the characteristic time. These expressions are valid for $|t| \to 0$. Further from T_C, terms representing noncritical contributions that vary smoothly through T_C can be included in fits to the critical behavior.

Another important critical exponent, ϕ governs crossover from one universality class to another as $|t| \to 0$. We will discuss its importance in the shape of phase boundaries and in the change in behavior that can occur as the asymptotic behavior at small $|t|$ changes as $|t|$ increases.

It has been assumed in the equations describing the asymptotic behaviors listed above that the critical exponents are the same above and below T_C. That is now believed to be generally true.

The universal critical exponents and amplitudes are not independent and there are scaling laws that relate them. Among the relationships that are generally true include

$$\alpha + 2\beta + \gamma \geq 2, \tag{2.8}$$

which is usually satisfied as an equality, and

$$\gamma = \beta(\delta - 1). \tag{2.9}$$

Another exponent, η, appearing in scattering line shape equations below, for example, is related to the exponents γ and ν by the equation

$$\gamma = \nu(2 - \eta). \tag{2.10}$$

The hyperscaling relation

$$d\nu = 2 - \alpha \tag{2.11}$$

includes the dimensionality d and does not always hold. For example, it is violated for dimensions above the critical dimension, $d \geq d_u$, where the critical behavior is described by mean-field theory. It is also violated in the random-field Ising model.

The ratios of amplitudes above and below T_C are also universal parameters. They are often measured with good accuracy in experiments, but they are not always measured as accurately in simulations or well predicted by theory. When amplitudes for different behaviors can be measured in the same set of experiments, the value of the combination referred to as two-scale universality (Bruce, 1981; Hohenberg *et al.*, 1976; Stauffer *et al.*, 1972),

$$R_S = \chi_0^+ (\kappa_0^+)^d / M_0^2, \qquad (2.12)$$

can be determined. This was measured in $d = 2$ Ising scattering experiments. As described in Section 4.7.1, the results (Cowley *et al.*, 1984a; Hagen and Paul, 1984) are reasonably consistent with the predictions.

The different behaviors that can be measured in experiments also have noncritical components that can vary depending on the experimental system. There may also be crossover behaviors, which need to be considered in fits of the data.

The exponents that have been determined from theory and simulations for some common universality classes are shown in Table 2.1 and amplitude ratios are given in Table 2.2. The variable d is the system dimensionality and the variable n specifies the degrees of freedom of the spins close to T_C. The Ising model corresponds to $n = 1$, the XY model is $n = 2$, and $n = 3$ is the Heisenberg model. The mean-field universality class holds at higher dimensions $d > d_u$ where fluctuations are not important. One-dimensional systems, with $d = 1$ are not shown because the thermal fluctuations are too large for the system to order. The same is true for $d = 2$ Heisenberg systems. The $d = 2$ XY model system undergoes the

Table 2.1: Critical exponents for several universality classes.

	MF	$d = 2, n = 1$	$d = 3, n = 1$	$d = 3, n = 3$
γ	1	1.75	1.2372(5)	1.39
ν	1/2	1	0.6301(4)	0.71
η	0	0.25	0.0364(5)	0.036
α	0	0 (log)	0.110(1)	−0.13
β	1/2	0.125	0.3265(3)	0.37
δ	3	15	4.789(2)	4.8

Notes: The values for the mean-field (MF) model and the $d = 2$ Ising model ($n = 1, d = 2$) are exact. The other values are taken from Pelissetto and Vicari (2002). The $n = 1$ Ising model exponents are estimates from all the tabulated results provided by the authors of the review. The $n = 3$ Heisenberg model are estimates based on the detailed tabulation by the authors. In addition to a large set of theoretical and simulation results, they also list a large number of experimental results.

Table 2.2: Amplitude ratios of some universality classes.

	MF	$d = 3, n = 1$	$d = 3, n = 3$	$d = 2, n = 1$
A^+/A^-	0	0.53–0.57	1.52(2)	1
χ_0^+/χ_0^-	1/2	4.8–5.0		37.7
ξ_0^+/ξ_0^-	1.95–2.06			3.16

Notes: The $d = 3$ values are estimates based on the detailed compilation in Pelissetto and Vicari (2002). The $d = 2$ and $n = 1$ values can be found in Delfino (1998).

Kosterlitz–Thouless transition (Kosterlitz and Thouless, 1973) with quasi-long-range order, which won't be covered here. The $d = 2$ Ising model has an exact solution (Onsager, 1944). The $d = 3$ universal parameters have been determined from theory and simulations. Although these are the most commonly encountered critical parameters and scaling relations, additional ones will be introduced as needed in the discussion of experiments exploring specific universal behaviors.

Scattering cross sections are given by the structure factor $S(\vec{q})$, which is the Fourier transform of the spin-spin correlation function. A general expression is

$$S(\vec{q}) = [\langle S_{\vec{q}} S_{-\vec{q}} \rangle], \tag{2.13}$$

where $\langle \ \rangle$ represents a thermal average and $[\ ]$ represents a configuration average that applies to systems with quenched randomness. The susceptibility can be expressed as

$$\chi(\vec{q}) = [\langle S_{\vec{q}} S_{-\vec{q}} \rangle - \langle S_{\vec{q}} \rangle \langle S_{-\vec{q}} \rangle]. \tag{2.14}$$

The structure factor can then be described as

$$S(\vec{q}) = \chi(\vec{q}) + [\langle S_{\vec{q}} \rangle \langle S_{-\vec{q}} \rangle] = \chi(\vec{q}) + \chi^{dis}(\vec{q}), \tag{2.15}$$

where $\chi^{dis}(\vec{q})$ is the disconnected susceptibility. For uniform systems, $\chi^{dis}(\vec{q}) = M^2 \delta(\vec{q})$ is the Bragg scattering intensity.

A major challenge in the analysis of scattering data is to have an accurate expression for $S(q)$. In mean-field theory, for a system with one relevant correlation length, we can write

$$S(q) = \frac{A}{q^2 + \kappa^2} + \frac{B}{(q^2 + \kappa^2)^2} + M^2 \delta(q), \tag{2.16}$$

where $B = 0$ for systems without quenched disorder and $M = 0$ for temperatures above the ordering temperature. The first two terms in this form are a Lorentzian and a squared-Lorentzian. To go beyond mean-field models, the asymptotic behaviors

$$\chi(q) \propto |q|^{-2+\eta}, \tag{2.17}$$

and

$$\chi^{dis}(q) \propto |q|^{-4+\bar{\eta}} \tag{2.18}$$

must be satisfied as the transition is approached. The mean-field values are $\eta = 0$ and $\bar{\eta} = 0$. More general, but still approximate, expressions then have the forms

$$\chi(q) = |q|^{-2+\eta}\tilde{\chi}(|q|/\kappa), \tag{2.19}$$

and

$$\chi^{dis}(q) = |q|^{-4+\bar{\eta}}\tilde{\chi}^{dis}(|q|/\kappa). \tag{2.20}$$

When two different correlation lengths are important near a transition, the equations can be easily modified to take into account the two values of κ.

Exact expressions for the neutron scattering line shapes are generally not available, so various approximations for $\tilde{\chi}(q/\kappa)$ and $\tilde{\chi}^{dis}(q/\kappa)$ are used for specific experiments, when available, and they will be described in detail as needed in the following chapters. When adequate expressions are not available, the accuracy of the experimental results can be limited.

In this book, the measurement of critical exponents and amplitudes obtained from fitting the data close to the transition will be emphasized because those measurements are the crucial tests of our understanding of the critical behavior of the systems. Results on model systems will be given and interpreted in the context of the best theoretical models available for data analyses.

To determine accurate values for the magnetic interaction strengths of a system, techniques such as inelastic neutron scattering and nuclear magnetic resonance (NMR) can be used. However, approximate values can often be obtained from magnetometry at temperatures well above any magnetic transition. For high enough temperatures, the inverse of the magnetic moment, $1/M$, in a small field H versus T will be nearly linear. The mean-field Curie–Weiss expression,

$$M/H = C/(T - \theta_{\mathrm{CW}}) + B, \qquad (2.21)$$

can then be used. The temperature θ_{CW} will be negative for a collinear anti-ferromagnet and positive for a ferromagnet and the absolute value should roughly approximate the transition temperature. The parameter C can be used to estimate the size of effective moments using

$$\mu_{\mathrm{eff}} = [3k_B C/\mu_B^2 N_A]^{1/2}, \qquad (2.22)$$

where μ_B is the Bohr magneton, k_B is Boltzmann's constant, and N_A is Avogadro's number.

Chapter 3

Background on Experimental Techniques

To adequately determine critical behavior close to a second-order phase transition, a general requirement is that the correlation length for fluctuations, ξ, should be large compared to all relevant magnetic interaction lengths. Insulating magnetic systems often have the advantage that relevant interactions between moments typically extend to only the nearest neighbors. In some cases, the temperature dependence of the correlation length has been measured, but many experiments are done without that explicit knowledge. In practice, it is often adequate to take measurements to reduced temperatures as small as 10^{-3} to get meaningful characterizations. In some experiments, reduced temperatures smaller than 10^{-4} have been achieved. At much smaller reduced temperatures, data are often limited by such things as sample temperature uniformity and crystal quality. Metallic systems can be problematic because the itinerant electrons can often act over distances large compared to lattice spacings, so the asymptotic critical behavior cannot easily be reached, making characterizations of asymptotic critical behavior difficult. Even insulating systems with short-ranged interactions can exhibit critical behavior that is not representative of the expected asymptotic universality class when there are significant crossover effects. Such systems require much smaller reduced temperatures to ensure that the measured behavior represents the asymptotic universal critical behavior. We will discuss such crossover in the MnF_2 and FeF_2 systems in Section 4.3.2. One practical test of asymptotic behavior is to analyze data taken over varying ranges of reduced temperature to see if the effective parameters vary. Only through extensive experimental studies can one be confident that the critical parameters of a particular universality class are adequately determined in a real system.

3.1   Thermometry

Central to the success of most experiments addressing critical behavior is the ability to approach the transition close enough for the measurements to meaningfully test the asymptotic universal parameters predicted by theory. There are practical considerations that can limit how closely the transition can be approached, both having to do with the experimental configuration and the quality of the crystal being studied.

The requirements for determining the absolute temperature in an experiment are less stringent than determining relative changes in temperature because the critical behaviors typically depend not on T, but on $t = (T - T_C)/T_C$, where T_C is the transition temperature. The universal behaviors become valid for data taken at small values of $|t|$. Different techniques have different requirements on the accuracy of the relative temperature measurements. For temperatures on the order of 10^2 K, temperature resolutions of 10 μK, or one part in 10^7 are typical of the best critical behavior studies when the data involve temperature differences, such as in measurements of the specific heat that require temperature changes of the sample resulting from small heat pulses. Other techniques that measure the temperature derivative of a quantity, such as the temperature derivative of the linear optical birefringence, $d(\Delta n)/dT$, require similar precision because the derivative must be accurately calculated for small temperature differences. Such temperature resolution requires careful attention to the thermometry details.

In addition to the thermometer resolution, the sample temperature must be uniform enough across the sample to allow meaningful data. Temperature variations, δT, should be small to ensure that $\delta T/T_C \ll |t|$. In pure crystals, sample mounting with good thermal contact to the sample holder and thermometry can be the limiting factor restricting the range of $|t|$ accessible in the experiment. Examples will be shown in the discussions in Section 4.3.2 of experiments on the specific heat of FeF_2 and MnF_2 that reach reduced temperatures less than 10^{-4}.

Scattering experiments are examples where such tight requirements on the thermometry can be relaxed slightly because each data point is taken at a fixed temperature. In such experiments, the relative temperature accuracy and stability need only to be sufficient to determine t accurately. A measurement at $t = 10^{-4}$ requires a relative accuracy to determine t of the order of 10^{-6}. An example of this is the neutron scattering for $T \approx 100$ K where a measurement near $|t| = 10^{-4}$ requires the relative temperature accuracy approaching 10^{-4} K. For the $d = 3$ anisotropic antiferromagnet FeF_2, with a transition temperature of 78 K, such accuracy was achieved, but that is an exceptional case.

Most experiments do not achieve reduced temperatures on the order of 10^{-4} but in many cases they can still adequately test theoretical results. Limitations sometimes come from the available apparatus and sometimes they are consequences of the sample quality. However, the comparison between theory and experimental results must then be done cautiously. In the case of FeF_2, we show in the next chapter that exponents can be measured reasonably well for $|t| > 10^{-3}$, but not the amplitude ratios. For the weakly anisotropic MnF_2, if data are only available for $|t| > 10^{-3}$, neither the exponent nor the amplitude ratio for the specific heat can be measured accurately because of crossover effects.

The control and measurement of the sample temperature requires isolation of the sample from the environment using heat shields. It also requires excellent thermal contact between the sample and thermometer, which often means thermally anchoring the sample and thermometer to a sample holder. For low temperature experiments, the sample can be in a vacuum, which is most stable, or a helium atmosphere, but flowing helium can prove to be unstable if the apparatus is not well designed.

Measuring the temperature accurately enough to reliably measure critical behavior requires the use of a thermometer that is well calibrated, has a measurable property with a smooth but significant variation with temperature, and is stable. For temperatures below room temperature, metallic resistance thermometry can provide smooth temperature versus resistance behaviors that can be easily calibrated. Nearly pure, unstrained copper wire can be used as an extraordinarily accurate thermometer when mounted on a copper sample holder. The resistance versus temperature behavior is universal in copper once the zero temperature residual resistance that comes from defects and impurities is measured. If wrapped around and secured to a copper sample holder, it will not be strained as the holder cools or heats. One disadvantage of copper wire is its significant magnetic field sensitivity. Another metal wire commonly used is made from platinum with a carefully controlled impurity level. The platinum wire is encapsulated and can be embedded in the copper sample holder using thermal grease. For specific heat experiments, metal elements can be evaporated onto sample platforms made of materials with small phonon specific heats, such as sapphire.

The significant field dependence of the resistance can be problematic with metallic wire thermometers. A good alternative is to use carbon-glass thermometers that have a small field sensitivity. The field dependence can be measured. The resistance temperature calibrations are very stable while cold and are fairly stable upon thermal cycling to room temperature. They can be purchased either encapsulated and embedded in the sample holder or they can be bare elements that can be attached to the sample holder or sample itself. The bare elements are useful in specific heat experiments when the mass of the thermometer must be kept to a minimum.

Accurate thermometry typically requires precision resistance bridges and four-wire resistance measurements to restrict the resistance measurement to the thermometer and not the leads; two wires are used to measure voltage across the thermometer while the other two are used to provide current. The current must be low enough to ensure it is not distorting the measurement by heating the thermometer or sample. Alternating current avoids spurious dc voltages from solder joints and strains in the wires. Low frequencies ensure accuracy of the resistance versus temperature calibrations. For the highest temperature sensitivity, bridges must be capable of high resolution resistance measurements and good stability. Resistance measurements accurate to one part in 10^8 are quite possible. The thermal anchoring of the leads to the sample holder is essential to remove uncontrolled heat leaks, particularly from the current leads.

Thermistors can be used to measure temperatures with less sensitive resistance bridges and can be obtained with extraordinary temperature sensitivity over limited temperature ranges. They tends not to be stable, especially when warmed to room temperature and then cooled. However, they can be calibrated against a stable, but less sensitive resistance thermometer after cooling. In this way, the sensitivity comes from the thermistor and the calibration comes from the resistance thermometer. Other forms of thermometry can be used as long as the required temperature sensitivity and stability are achieved and the measured property of the thermometer is smooth enough to preclude introducing systematic noise into the critical behavior data.

Whatever thermometer is used, it must not generate excessive heat, especially in specific heat measurements. It must also have its leads well anchored to the sample holder and to the cryostat shields so that the leads do not leak excessive heat to the sample.

3.2   Specific Heat Techniques

Direct specific heat techniques can probe the magnetic specific heat critical behavior. However, to obtain good results, great effort must go into

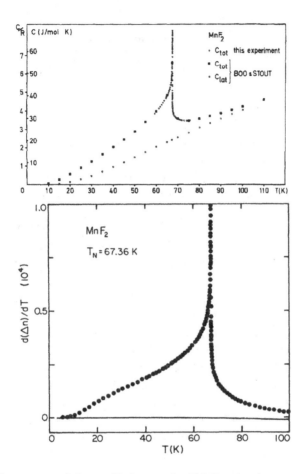

Fig. 3.1: Measurement of the specific heat peak of MnF$_2$ using (upper panel) a classic pulsed heat technique (Nordblad *et al.*, 1981) and (lower panel) the optical birefringence technique (Belanger *et al.*, 1983b). The nonmagnetic contributions are dramatically reduced by using the optical measurements. Upper figure is reproduced with permission from Elsevier.

ensuring accurate measurements. The finest measurements use a pulsed heat source of some kind and precision thermometry, as discussed in the previous section. Extraordinarily fine measurements on MnF$_2$ (Nordblad *et al.*, 1981) are shown in the upper panel of Fig. 3.1. We will discuss the critical behavior analysis later, but what is clearly present is the large specific heat contribution from phonons. Fortunately, the phonon contributions had been previously determined (Boo and Stout, 1976) so that they could be subtracted with good accuracy. Also shown in Fig. 3.1, in the lower panel, is the magnetic specific heat determined using the temperature derivative of the optical birefringence (Belanger *et al.*, 1983b), a technique that will be discussed in the next section. The non-magnetic contributions are small and can be subtracted. The comparison of the critical behaviors will be covered in Section 4.3.2.

The situation for the specific heat versus T of the $d = 2$ Ising model is much more challenging with direct specific heat measurements. The upper panel of Fig. 3.2 shows the measurements (Ikeda *et al.*, 1975) using K_2CoF_4 and demonstrates how the phonon contribution dominates the measurements to the point of limiting the critical behavior characterizations. Note that the vertical axis does not start at zero. The nonmagnetic background of the $d(\Delta)/dT$ versus T data taken (Nordblad *et al.*, 1983b) on the sister compound Rb_2CoF_4, shown in the lower panel of Fig. 3.2, is extremely small in comparison, allowing a much more accurate determination of the critical behavior, as discussed in Section 4.3.1. Sometimes, the direct specific heat technique is not the best way to determine accurate critical behavior of the magnetic specific heat.

Typically, but not always, magnetic systems equilibrate on time scales short compared to measurement times, so a primary concern with sample equilibrium is that the temperature remains uniform across the sample as the energy is applied and this requires good thermal conductivity within the sample. Powdered samples can be quite difficult to equilibrate. The sample must be in excellent thermal contact with the thermometer so the measured temperature is truly that of the sample. For specific heat measurements, the wires used for thermometry, heaters, and sample support must be carefully planned. Fortunately, high quality wires of very small diameter are available, including both pure copper and gold, as well as various alloys that provide a range of thermal and electrical conductivity. High electrical conductivity reduces the heating in the wires while low thermal conductivity helps with sample isolation. A combination of wires are used as heaters and leads to the thermometer. The mounting of samples can be important. Silver paint is often used because it provides good thermal contact between the sample and holder without straining the sample. Thermal grease that does not harden at low temperatures is also useful. A weak spring can be used to hold the sample in place. However, for birefringence experiments, unintended strain that produces unwanted birefringence must be avoided.

Among the most challenging with respect to thermometry are pulsed heat specific heat experiments. In pulsed heat specific heat techniques,

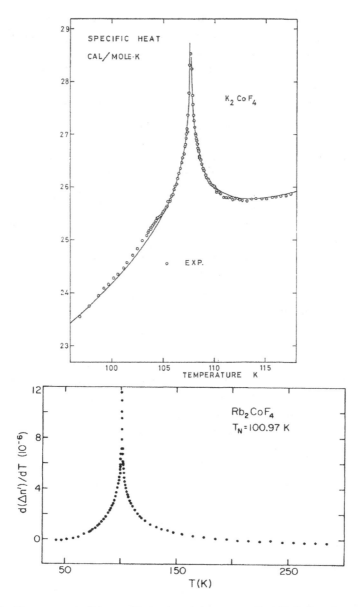

Fig. 3.2: Measurement of the specific heat peak (upper panel) of K_2CoF_4 using a classic specific heat technique (Ikeda *et al.*, 1975) and (lower panel) Rb_2CoF_4 using the optical birefringence technique (Nordblad *et al.*, 1983b). The nonmagnetic contributions are dramatically reduced by using the optical measurements. Note that the scale in the upper panel does not start at zero; the phonon specific heat is a large part of the signal. Upper figure is reproduced with permission from the Physical Society of Japan.

the sample is thermally isolated, energy is applied in some form, and the resulting temperature increase is inversely related to the specific heat. Implicit assumptions for an accurate equilibrium critical behavior measurement are that the system is sufficiently equilibrated during the measurements, the increase in temperature is small relative to the difference between the system's temperature and the transition temperature, and the sample is uniform with respect to physical properties relevant to the transition. An experiment must be designed so that the temperature change resulting from the externally applied energy source is significant compared to drift in temperature without the external source. The average sample temperature might be stable, but if heat is entering on part of the sample and leaving another part, the two parts will have temperatures that are not same distance from the transition temperature; sample mounting and shielding are particularly important in this regard. As the transition is approached in temperature, the size of the energy pulses applied to the sample must be small enough so that the measurement is not averaged over a rapidly evolving specific heat. In mixed magnetic systems where the transition temperature is a function of the concentration, the uniformity of the sample is often a limiting factor in obtaining accurate critical behavior measurements.

Often, specific heat is measured using techniques in which the sample temperature is varying significantly. The specific heat is then inferred from the rate at which the sample and its holder are changing in time. Although good accuracy can be achieved, equilibration issues are important to consider. For the random-field Ising model experiments to be discussed later, the relaxation of the magnetic system is extremely slow, so that experiments that are often considered to be static on normal laboratory time scales are in fact not. In this case, the technique used to measure specific heat must be carefully designed. Pulsed heat techniques might be preferred in this case.

The possible variations in specific heat techniques are numerous and it would be impossible to review them all here. However, a technique we used to measure the specific heat in large magnetic fields might be of use in some cases. The design challenge is to have precise temperature control that prevents overshooting the target temperature, while fitting within the confined space of the central bore of a superconducting magnet. For the typical temperatures $30 < T < 100$ K used in the experiments described in this book, temperatures were measured with a relative accuracy much better than $50\,\mu$K. The small field dependence of the carbon–glass thermometers was measured using the thermally isolated sample thermometer, measuring in zero field, measuring the resistance again after raising the field, and measuring once more after zeroing the field. The changing of the field slightly heats the sample holder, but measuring in zero field before and after allows compensation for that effect since the same heating occurs when the field increases or decreases. This process is repeated for temperatures over the range of specific heat measurements. The sample is mounted with silver paint on a sapphire plate along with two bare carbon-glass thermometers and a heater. One calibrated thermometer is used to measure the sample temperature using a low frequency ac bridge with an eight-decade transformer, stable calibrated reference resistors and lock-in amplifier detection. The heater is used to deliver a timed heat pulse using a constant current source. Four leads are used for the heater. Two deliver the current and two monitor the voltage. An electronic circuit ensures that the current times the voltage is constant and is applied for a specified time period. The second thermometer on the sapphire plate is used with another low frequency ac bridge that is identical to the first except that, in place of a reference resistor, the bridge uses a third thermometer placed on the sample shield. The sample shield has a control heater and is connected to the cryostat cold finger using a narrow copper neck that provides a weak thermal connection. The cold finger of the cryostat has an additional copper shield that surrounds the sample and sample shield. The second bridge is used to control the difference in temperature between the sapphire sample plate and the copper sample shield. All wires near the sample are twisted pairs and are thermally anchored to the thermal shields. Inside the cryostat, miniature coax cables are used away from the immediate region of the sample. The sample and shield are kept in a vacuum and the various wires are chosen to control the heat flow between the sample and shield so that the sample drift can be nulled. Before the specific heat of the sample is measured, the small temperature difference between the shield and sample is calibrated to eliminate sample temperature drift. The sample temperature is then set, a tiny heat pulse is applied for about 30 seconds and, as the sample temperature changes, the calibrated difference between the sample and shield is adjusted to ensure that the only sample temperature change is from the heat pulse. The heat pulse measurement is controlled by a computer program that uses the thermometer and sample-shield calibrations. The design allows fine pulsed heat measurements within the confines of the 7 T superconducting solenoid. The fine temperature control is essential for critical behavior studies, particularly for the random-field Ising experiments where

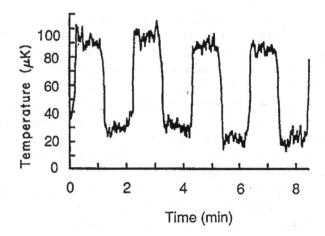

Fig. 3.3: The stability of the thermometry used in the pulsed heat specific heat measurements described in the text (Dow and Belanger, 1989). The jumps at one minute intervals shows the resolution of the thermometry and the overall drift over eight minutes demonstrates the stability of the system.

the temperature can not be allowed to overshoot the target temperature (see Section 4.4).

To demonstrate the temperature resolution and stability of the technique described in the previous paragraph, the apparatus near 50 K was stabilized to minimize drift. As shown in Fig. 3.3, the sample thermometry bridge setting was changed every minute by a small amount amount to show the resolution of the temperature measurement. The sample drifted in temperature by a small amount, less that $20 \, \mu K$ over the eight minutes shown in the figure.

3.3 Optical Measurements of Critical Behavior

Optical birefringence, Faraday rotation, and the related capacitance techniques have played a key role in a number of experiments on anisotropic and isotropic antiferromagnets, but they are not widely used. The study of antiferromagnets with quenched nonmagnetic impurities have benefited from birefringence experiments, as will be discussed in some detail. However, the results of optical studies, in particular birefringence measurements, have been questioned in the literature (Birgeneau *et al.*, 1996; Hill *et al.*, 1997; Wong *et al.*, 1982). It is therefore worthwhile to describe the techniques in some detail so that the experimental results can be fully appreciated.

3.3.1 *Optical birefringence*

The birefringence was measured using the Sénarmont technique depicted in Fig. 3.4. A polarized laser beam with a wavelength not strongly absorbed by the crystal is aligned with the polarization direction at precisely 45° to the two optical axes of the crystal. The light after the crystal is elliptically polarized because the polarizations interact with the different indices of refraction along the two optical axes. A quarter-wave plate, particular to the laser wavelength, is used to transform the elliptically polarized light into linearly polarized light at an angle proportional to the birefringence Δn. A quartz modulator adds a small oscillating signal that allows the relative Δn to be determined to one part in 10^8, which is a small part of the typical birefringence of anisotropic insulators, as illustrated in Table 3.1. Precise alignment with the sample to the optics is required, but easily achieved (Belanger, 1981). The output of the detector will show a component that

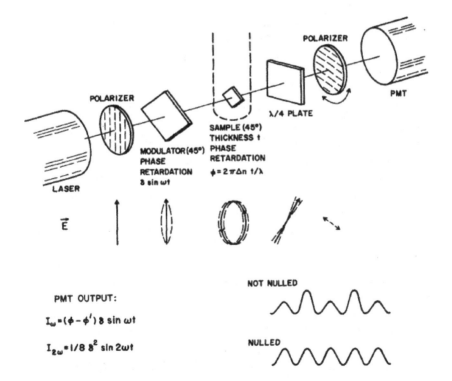

Fig. 3.4: The Sénarmont technique (Belanger, 1981; Belanger *et al.*, 1984) for measuring the optical linear birefringence.

Table 3.1: The index of refraction and magnitude of the optical birefringence (Jahn, 1973).

	FeF_2	MnF_2	ZnF_2
n	1.516	1.485	1.508
Δn	0.0100	0.0286	0.0292

has the frequency of the modulator, ω, and another at a frequency 2ω. When the optics are well nulled, the analyzing polarizer can be rotated until the detector output has only the 2ω component. As the birefringence changes with temperature, the lock-in amplifier's ω output signal is minimized as the analyzing polarizer tracks the changing polarization angle. The analyzing polarizer is controlled by the computer using a stepper motor and the output of the lock-in amplifier. Low birefringence fused quartz windows are mounted in a strain-free manner on the room-temperature cryostat access ports. Inside the cryostats, the laser beam travels through the vacuum, with small holes in the shields to allow access to the crystal with minimal heating from thermal radiation. With high resolution thermometry, $d(\Delta n)/dT$ versus T can be measured accurately.

The birefringence of antiferromagnetic crystals can change near magnetic phase transitions and, for some crystals, this presents opportunities to determine critical behaviors. Fortuitously, the technique works well in key systems, particularly the fluoride antiferromagnets covered in Chapter 4. The interpretation of the data is different for anisotropic and isotropic systems. In magnetically anisotropic systems, the crystal remains optically anisotropic through the magnetic transition. In that case, $d(\Delta n)/dT$ is proportional to the magnetic specific heat critical behavior (Jahn *et al.*, 1977). In isotropic systems, the birefringence measurement is a probe of the breaking of symmetry at a magnetic transition (Gehring, 1977). Cubic crystals above the magnetic transition have no birefringence, but the magnetic ordering below the transition temperature breaks the magnetic and optical symmetry.

A key to the effectiveness of the technique is the resolution of the measurement. Using modulation and lock-in detection, the changes in the birefringence, Δn, can be measured with a resolution of one part in 10^8. This does require the apparatus itself to be free of birefringence. The sample also must be mounted without stressing it in unintended ways. A key feature of the birefringence measurements that allows high sensitivity is that one polarized laser beam is used to simultaneously measure the difference in the index of refraction in two directions using the two polarizations of light, one along each of the two optical directions of the crystal. The path through the sample is identical for the two polarizations of the single laser beam. Effects from vibrations are the same for both polarizations and cancel. Exactly the same part of the crystal is probed by both polarizations. Another advantage of the technique is that only a small part of the sample is probed and, for mixed crystals, the path through the crystal can be chosen to be approximately perpendicular to the concentration gradients. Experimental details can be found in Belanger (1981) and Belanger *et al.* (1984).

The history, theory and application of the optical linear birefringence technique to magnetic systems has been reviewed in great detail (Ferre and Gehring, 1984) for anisotropic and isotropic magnetic crystals. The birefringence in magnetic crystals has a number of possible sources, and it is hard in most cases to know precisely what the mechanisms are for a given crystal. However, knowing the sources is not essential as they all lead to the same behaviors.

3.3.2 *Specific heat critical behavior via birefringence techniques*

The proportionality between the magnetic contributions to the specific heat and $d(\Delta)/dT$ is particularly important to discuss because of its role in several key experiments in Chapter 4. In anisotropic crystals, the magnetic contribution to the specific heat is given by the temperature derivative of the various spin–spin correlations and the interaction strengths between each set of spins. Regardless of the sources of the birefringence, the magnetic contribution to the temperature dependence of the birefringence is also related to the temperature dependence of the various spin–spin correlations, but the contribution from each set of spins is not necessarily related to the interaction strengths. The proportionality between the magnetic parts of the specific heat and $d(\Delta n)/dT$ only holds if there is only one dominant spin–spin correlation governing both. Otherwise, a term proportional to the temperature derivative of the square of the order parameter could possibly dominate below the transition (Ferre and Gehring, 1984). Such a term has not been observed in the fluorides and an applied field would not change that. If only one interaction is important in the specific heat and $d(\Delta n)/dT$, the proportionality should then hold. This must be the case in the fluorides, as shown in detail for MnF_2 and FeF_2 below.

The proportionality of the specific heat and $d(\Delta n)/dT$ has been tested in MnF_2 over a wide range of temperatures after correcting for the large phonon background specific heat in pulsed heat measurements and the small background in $d(\Delta n)/dT$ measurements. The two can be normalized by integrating the entropy for $0 < T < 550$ K; the entropy for $T > 550$ K is negligible. The proportionality between the specific heat and $d(\Delta n)/dT$ then holds well over the entire range of temperatures (Belanger *et al.*, 1984). If the magnetic specific heat, C_m, and the magnetic contribution of the birefringence derivative, $d(\Delta n)_m/dT$, are related by a proportionality constant, it can be defined by

$$C_m/Nk_B = (d(\Delta n)_m/dT)/K, \qquad (3.1)$$

where N is Avogadro's number and k_B is Boltzmann's constant. The entropy can then be calculated accurately as

$$\ln(6) = \int_0^{550\text{K}} (1/KT)d(\Delta n)_m/dT, \qquad (3.2)$$

where the factor 6 comes from $2S + 1$, with $S = 5/2$ being the spin of the Mn ions in MnF_2. The proportionality holds extremely well over the entire temperature range $0 < T < 550$ K, including near the transition temperature.

The proportionality also holds well in the magnetically dilute $Fe_xZn_{1-x}F_2$. Figure 3.5 shows $d(\Delta n)_m/dT$ versus T for several magnetic concentrations after subtracting the phonon background (Ferreira *et al.*, 1991a). The critical behavior decreases rapidly with dilution and the noncritical component becomes dominant. At the magnetic percolation threshold concentration, x_p, the noncritical contribution is, of course, the only part. If $d(\Delta n)_m/dT$ is proportional to C_m, the change in energy should also be proportional to the change in $(\Delta n)_m$. Since the energy should vary as the number of interacting bonds, it should correspondingly vary as x^2. Figure 3.6 shows the change in $(\Delta n)_m$ versus x^2, and the dependence is represented by the straight line. Hence, the proportionality between $(\Delta n)_m$ and energy holds even upon dilution. A similar study was made for diluted MnF_2, as shown in Fig. 3.7, and the same linear relationship between $(\Delta n)_m$ and the energy is established (Jahn *et al.*, 1977).

Measuring specific heat using heat pulses is not necessarily the most effective technique for magnetic crystals. This is the case, for example, when the nonmagnetic contributions are large or when there are significant concentration gradients. Optical birefringence can be effective when the

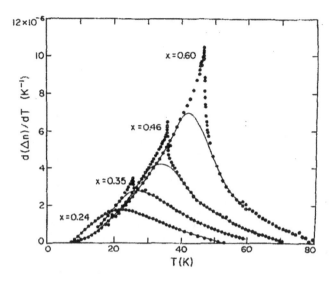

Fig. 3.5: The magnetic part of $d(\Delta n)/dT$ versus T for $Fe_xZn_{1-x}F_2$ at several magnetic concentrations x (Ferreira *et al.*, 1991a). The critical contribution decreases with x and the noncritical part dominates at x approaches the magnetic percolation threshold concentration $x_p = 0.24$. The solid curves are estimations of the noncritical magnetic contributions. The overall change in Δn varies with x^2, as the magnetic energy also should. Reproduced with permission from AIP Publishing.

crystal exhibits a measurable $\Delta(n)$ versus T, but that is not always the case. It has proven especially effective for the fluorides such as $Fe_xZn_{1-x}F_2$, $Mn_xZn_{1-x}F_2$, and $Fe_xMn_{1-x}F_2$. On the other hand, some systems do not show a birefringence effect large enough to be useful. This is, unfortunately, true (Kato *et al.*, 1992) for the triangular lattice systems such as $CsMnBr_3$ covered in Chapter 6.

Optical birefringence has two other advantages for magnetically dilute crystals. First, the concentration variations in some mixed crystals can be measured throughout the crystals at room temperature, and the effects of a gradient on the measurements can then be minimized if the laser beam can be oriented to be perpendicular to the direction with the greatest variation in concentration, as demonstrated in Section 3.7. Second, magnetic dilution decreases the magnetic specific heat critical peak relative to phonon contributions and noncritical magnetic contributions grow significantly. Thus, many studies of the random-exchange and random-field specific heat measurements rely on the optical birefringence technique. An early example (Belanger *et al.*, 1983a) of the specific heat characterization in $Fe_{0.6}Zn_{0.4}F_2$

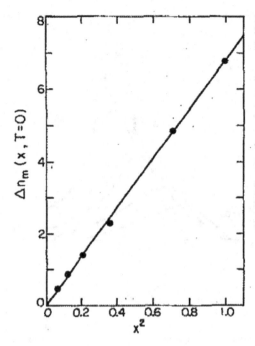

Fig. 3.6: The change in the magnetic contribution to Δn versus x^2 for $Fe_x Zn_{1-x} F_2$ using the data from Fig. 3.5 demonstrating the proportionality of the magnetic contributions to the specific heat and temperature derivative of the birefringence (Ferreira *et al.*, 1991a). Reproduced with permission from AIP Publishing.

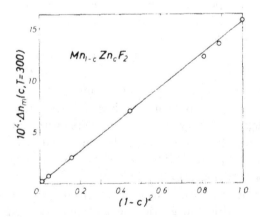

Fig. 3.7: The change in the magnetic contribution to Δn versus $(1 - c)^2 = x^2$ for $Mn_{1-c} Zn_c F_2$ demonstrating the proportionality between the magnetic contributions to the specific heat and temperature derivative of the birefringence (Jahn *et al.*, 1977). Reproduced with permission from Elsevier.

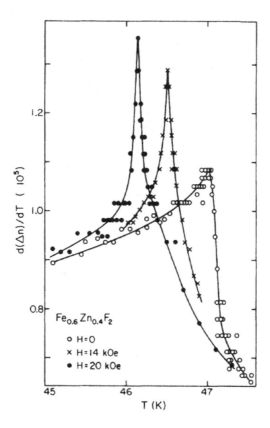

Fig. 3.8: Birefringence measurements of the magnetic specific heat in $Fe_{0.6}Zn_{0.4}F_2$ (Belanger *et al.*, 1983a).

using $d(\Delta)/dT$ versus T is shown in Fig. 3.8. It shows the dramatic influence of a small applied field on the magnetic specific heat critical behavior as the system crosses over from $d = 3$ random-exchange to random-field Ising behavior.

Just as with the pure magnetic systems, birefringence ($d(\Delta n)/dT$) data (Ferreira *et al.*, 1991b) exhibit the same magnetic behavior as the direct specific heat data for a $Fe_{0.46}Zn_{0.54}F_2$ crystal in zero and non-zero fields applied along the unique axis. The specific heat, measured using a pulsed heat technique (Dow and Belanger, 1989), and is compared with $d\Delta n/dT$ measurements (Ferreira *et al.*, 1991b) in Fig. 3.9. An important observation is that both the specific heat and the temperature derivative of the birefringence data versus T show a more rounded peak when the crystal is cooled

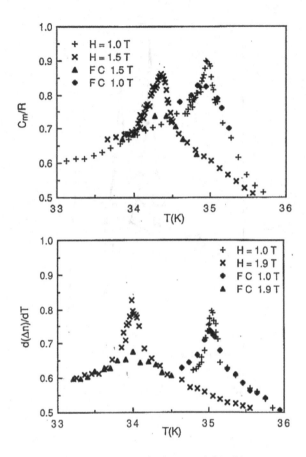

Fig. 3.9: Comparison of magnetic specific heat and birefringence measurements in $Fe_{0.46}Zn_{0.54}F_2$ (Dow and Belanger, 1989). The peak is small compared to the previous figure. The FC-ZFC hysteresis is observed with both techniques. The birefringence data were kindly provided by I. B. Ferreira.

in a field (FC) than when the crystal is cooled in zero field before taking data while heating in the applied field (ZFC). However, that difference is much easier to observe optically for the reasons described above. To see the subtle difference between the FC and ZFC data using pulsed-heat techniques, a sample with an extraordinarily uniform concentration is required. The concentration of one sample that shows the difference, $Fe_{0.46}Zn_{0.54}F_2$, was measured (King *et al.*, 1988) to have a change of concentration of only 2×10^{-4} across the entire crystal. The birefringence technique can minimize the effect of the gradient more than techniques such as direct specific heat that require the whole sample. Indeed, the direct specific heat peak is more rounded than the one shown by $d(\Delta n)/dT$ versus T. The difference between ZFC and FC data for $H = 1.5$T occurs over a small temperature region approximately $10^{-2} \times T_C(H)$, so the extraordinary quality of the $Fe_{0.46}Zn_{0.54}F_2$ was needed to see it clearly.

Later, FC specific heat data measured using a $x = 0.5$ sample did not appear to be different from ZFC (Birgeneau *et al.*, 1996; Hill *et al.*, 1997), seemingly contradicting the earlier study cited above. It was speculated that the birefringence sees a contribution from long-range order, but that would contradict all the experimental evidence and theory (Belanger *et al.*, 1996; Ferre and Gehring, 1984). It would also render random-field Ising scaling of the $d = 2$ and $d = 3$ data for $d(\Delta n)/dT$ as a function of field impossible, but that scaling is shown in Chapter 4. More likely, the concentration variation caused rounding of the peak in the $x = 0.5$ experiments (Birgeneau *et al.*, 1996; Hill *et al.*, 1997) that obscured the additional rounding of the FC peak versus ZFC peak. The random-field transition varies with the field (Eq. 4.22), so the rounding exhibited in a sample with $H = 0$ is amplified as H increases, further obscuring the difference. There is no theory or experiment that has demonstrated that magnetic part of $d(\Delta n)/dT$ is not proportional to the magnetic specific heat in the fluoride systems. On the contrary, the evidence for the proportionality is overwhelming.

The birefringence technique has some limits. It requires optical quality crystals that are transparent to the laser light. The temperature dependence of the birefringence signal also needs to be suitably strong.

3.3.3 *Faraday rotation*

Faraday rotation is an optical technique for measuring the net moment in an antiferromagnetic magnetic system with an applied field. The rotation of the polarization of the laser beam is proportional to the net moment, so it can be an ideal way to measure critical behavior. The laser beam and the field are directed precisely along the unique axis of the antiferromagnetic crystal. It is important to align the beam carefully to avoid a birefringence contribution to the signal. In practice, an alignment within $1°$ is adequate as the effect from a misalignment angle δ tends to grow as $\sin^2(\delta)$ (Belanger, 1981). An example of the linear and nonlinear susceptibility, measured with Faraday rotation, is shown for $FeCl_2$ in Fig. 3.10. The Faraday rotation was used to investigate random-exchange and random-field critical behavior

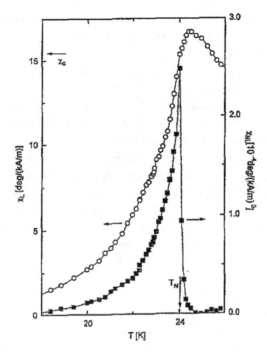

Fig. 3.10: The linear (χ_L) and nonlinear (χ_{NL}) susceptibilities of FeCl$_2$ (Kushauer and Kleemann, 1995). Reproduced with permission from, IOP Publishing.

in dilute antiferromagnets and it was shown (Kleemann *et al.*, 1986), by examining the scaling of the free energy, that

$$F(T, H) \propto H^{2(\alpha^{\mathrm{rf}} - \alpha)/\phi} |t|^{2 - \alpha^{\mathrm{rf}}}, \tag{3.3}$$

that

$$(\partial M / \partial T)_H \propto H^{(2/\phi)(1 + \alpha^{\mathrm{rf}} - \alpha - \phi/2)} |t|^{-\alpha^{\mathrm{rf}}}, \tag{3.4}$$

and

$$(\partial M / \partial H)_T \propto H^{(2/\phi)(2 + \alpha^{\mathrm{rf}} - \alpha - \phi)} |t|^{-\alpha^{\mathrm{rf}}}, \tag{3.5}$$

where α^{rf} is the random-field specific heat exponent, α is the random-exchange ($H = 0$) exponent, and ϕ is the crossover exponent. Magnetometry can also be used to measure the same properties of an antiferromagnetic system.

Faraday rotation is sensitive to the net moment from domain walls at low temperatures and has been used to study domain wall dynamics.

The time dependence of metastable domains generated by random fields has been studied for $H = 0$ using Faraday rotation as well as magnetometry techniques, as discussed in Section 5.1.

3.3.4 *Isotropic magnets and the birefringence technique*

For Heisenberg systems, the birefringence above the transition is zero because the system has cubic symmetry. Below the transition, the birefringence signal is non-zero within a domain, but cancels macroscopically if many domains are sampled along the beam path. The change in birefringence for temperatures below the transition temperature can be measured if single domains can be observed. A macroscopic birefringence can also be observed if the domains are oriented by a small uniaxial pressure or a small magnetic field. The Heisenberg nature of the transition is preserved if the pressure or field strengths are large enough to align the domains along a preferred direction but small enough to have negligible effect on measurable critical behavior. The birefringence signal grows as the temperature decreases.

The signal below the transition is predicted to be governed by the Heisenberg to Ising crossover exponent (Ferre and Gehring, 1984; Gehring, 1977), reflecting the breaking of the cubic symmetry by the ordering of the magnetic moments, with

$$\Delta n \propto |t|^{2 - \phi - \alpha_H}, \tag{3.6}$$

where α_H is the Heisenberg specific heat exponent and ϕ is the crossover exponent.

3.4 Capacitance

It is not surprising that crystals showing optical effects near the antiferromagnetic phase transition would also exhibit critical effects in the capacitance for $H = 0$, which is sensitive to the zero frequency dielectric properties (King *et al.*, 1984). In addition to the dielectric properties, crystal expansion, which also shows the specific heat critical behavior for $H = 0$, can contribute to the temperature dependence of the capacitance. With a properly shielded apparatus, the temperature dependence of the capacitance, dC/dT, versus T in zero applied fields rivals other techniques in anisotropic crystals, as demonstrated in Fig. 3.11 with dC/dT versus T showing the magnetic specific heat critical behavior of FeF_2. In addition, for $H = 0$, the specific heat critical behavior of isotropic systems such as $KNiF_3$ can be

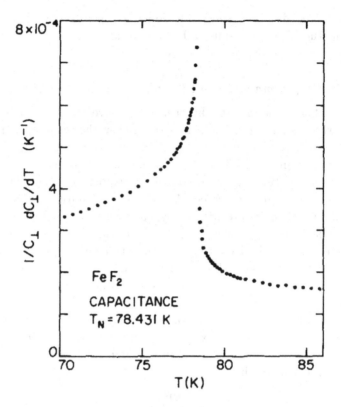

Fig. 3.11: Capacitance measurement of the $H = 0$ specific heat of FeF$_2$ (King *et al.*, 1984).

measured using this technique and compares well with traditional specific heat techniques, as will be shown in Section 4.9. For $H \neq 0$, the proportionality breaks down, as will be discussed in Chapter 4. On the other hand, it is very useful in the study of hysteresis in the random-field Ising model experiments.

3.5 Susceptibility and Thermal Expansion

Ideally, the parallel magnetic susceptibility and magnetic specific heat of an antiferromagnet are related (Fisher, 1962) by

$$C_M(T) \simeq A \frac{\partial}{\partial T}(T \chi_{\parallel}(T)). \tag{3.7}$$

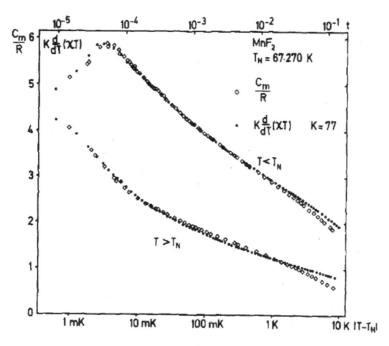

Fig. 3.12: The specific heat and susceptibility for $H = 0$ of MnF_2 (Nordblad *et al.*, 1981). Reproduced with permission from Elsevier.

However, although the relationship holds quite well for MnF_2 (Nordblad *et al.*, 1981) as shown in Fig. 3.12, the relationship fails for FeF_2 (Chirwa *et al.*, 1980); the discrepancy in the latter case is attributed to a small spontaneous moment that is approximately field independent. The spontaneous moment, resulting from magnetostriction, was used (Mattsson *et al.*, 1994) to obtain a value of the Ising order parameter critical exponent β that agrees with the value obtained using Mössbauer techniques (Wertheim and Buchanan, 1967) (see Section 4.4).

The thermal expansion was also measured close to T_N in MnF_2 (Nordblad *et al.*, 1981) as shown in Fig. 3.13. The measurements required a relative resolution of 2×10^{-9} of the sample length. Again, there is close agreement in the critical behavior of the expansion and the specific heat. In FeF_2 and $Fe_x Zn_{1-x} F_2$, it can be used to measure the order parameter in an applied field (Ramos *et al.*, 1988b), again because of piezomagnetic effects.

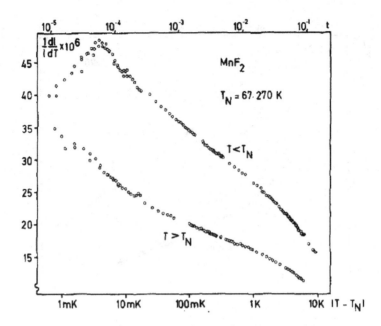

Fig. 3.13: The thermal expansion at $H = 0$ of MnF_2 (Nordblad *et al.*, 1981). Reproduced with permission from Elsevier.

3.6   Neutron and X-ray Scattering Measurements

The spectrometers used in neutron and X-ray scattering experiments vary greatly in design to optimize particular measurements. Figure 3.14 depicts the basic components of a triple-axis spectrometer typically used in neutron scattering critical behavior experiments. The source can be a thermal nuclear reactor or neutron spallation source. The monochromator is a natural or manufactured crystal used to select one wavelength using a Bragg reflection. Filters may be used to eliminate high energy neutrons that might also scatter at the selected angle. The various collimators help to define the beam. The beam scatters from the sample crystal and the scattered neutrons either go directly into the detector for elastic measurements, or scatter from an additional analyzer crystal if the scattered neutron energy is to be determined.

For measurements of the scattering near a ferromagnetic Bragg peak, where the magnetic and nuclear scattering peaks coincide, polarized neutrons can be used to distinguish between the two contributions by detecting whether the scattered neutron spin flips in a magnetic scattering process

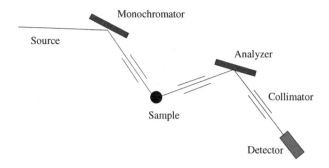

Fig. 3.14: A typical triple-axis spectrometer used in neutron scattering.

or does not flip in a nuclear scattering process. Antiferromagnetic scattering peaks do not generally coincide with the nuclear peaks. For example, it is typical for the unit cell to double in size when the sample orders antiferromagnetically, resulting in antiferromagnetic Bragg peaks at distances half way between the nuclear Bragg peaks in reciprocal space.

Synchrotron scattering can also be used for elastic scattering in antiferromagnets if the X-ray energy is chosen so that the region of reciprocal space near the antiferromagnetic Bragg peak is free of spurious atomic scattering.

3.6.1 *Elastic scattering*

The order parameter of magnetic systems undergoing phase transitions can be determined from the Bragg scattering intensity as a function of temperature at points in reciprocal space that correspond to the symmetry of the magnetic order. For ferromagnets, this can be problematic because the magnetic Bragg points coincide with lattice nuclear scattering points. Antiferromagnets have a distinct advantage in that the magnetic order scattering points do not coincide with the nuclear ones. Although it might seem that neutron scattering is the ideal technique to determine the order parameter of an antiferromagnet, and often is, the temperature dependence of the scattering amplitude can be altered by extinction effects in large, high quality crystals, particularly at $d = 3$ Bragg scattering points. The intensities in such crystals tend to saturate, distorting the temperature dependence of the intensities. In this case, X-ray scattering can be employed. The magnetic scattering cross section of X-rays is a few orders of magnitude smaller than electron scattering. Nevertheless, at the antiferromagnetic Bragg scattering point, there is no scattering from electrons.

The overall electron scattering is so strong that the beam is depleted within a few microns and this is not temperature dependent. The extinction effect thereby provides a fixed volume from which X-rays can weakly scatter magnetically. The intensity of the magnetically scattered X-rays from this volume is not affected by extinction. In this way, the temperature dependence of the antiferromagnetic order parameter can be accurately measured using the X-ray technique.

In addition to Bragg scattering for $q = 0$, the fluctuation correlation length and staggered susceptibility of an antiferromagnet can be measured using neutron scattering for $q \neq 0$. Synchrotron X-ray scattering can also be useful, but typically it is better for fluctuations on a very long length scale and for Bragg scattering. In either case, the resolution of the instruments must be accounted for in analyses of the data.

3.6.2 *Inelastic scattering*

Inelastic scattering, where the neutron changes energy in the scattering process, can be accomplished with the analyzer crystal. Dynamic critical behavior can be measured using the neutron spin-echo technique.

3.7 The Effect of Concentration Gradients on Critical Behavior Characterizations

Some of the interesting studies of phase transitions involve crystals grown as mixed magnetic systems or mixed magnetic and nonmagnetic (diamagnetic) systems. The experimental research into random-exchange and random-field effects provides crucial results that impact the relevant theoretical research. Research in such areas depends on minimizing, measuring and accounting for the effects from gradients. When the transition temperature or other properties vary with the concentration in a system being studied, it is essential to use crystals that are as uniform in concentration of the constituents as possible. In practice, it is difficult to avoid gradients that limit the critical behavior measurements, though they have been minimized in particular cases. To accomplish that, a great deal of attention was focused on measuring sample gradients and producing the highest quality crystals for experiments. There is much danger in interpreting experiments in dilute magnetic systems without taking concentration gradients into account; the physical properties of the system must not be confused with the effects of gradients on the experimental measurements.

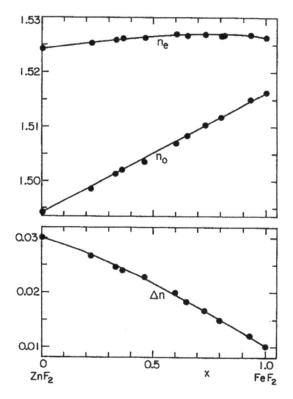

Fig. 3.15: The indices of refraction and birefringence of $Fe_xZn_{1-x}F_2$. The strong variation with x allows the concentration variations in a crystal to be measured (King *et al.*, 1988).

For the fluorides used in studies of random-exchange and random-field experiments, optical measurements were found to be an an accurate tool to characterize gradients in the most studied case of $Fe_xZn_{1-x}F_2$, partly because the birefringence is significantly different for FeF_2 and ZnF_2, as shown in Fig. 3.15. As the crystals grow from the melt, the boules tend to add the two constituents at slightly different and varying rates, resulting in concentration variations, particularly along the boule axis. After polishing parallel faces along the boule, the laser beam is scanned along the boule axis to measure the concentration changes. The measured concentration variations allow the most uniform sections to be cut from the boules for experiments. The faces of the cut boule sections can then be polished and the typically smaller gradients can be measured across the diameter. To demonstrate the effectiveness of this technique, optical birefringence

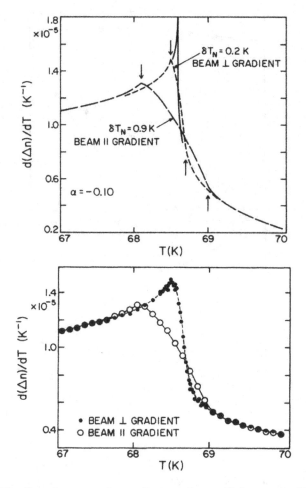

Fig. 3.16: The effect of concentration gradients on the specific heat peak of $Fe_{0.8}Zn_{0.2}F_2$ (Belanger *et al.*, 1988a; King *et al.*, 1988). In the upper figure, the gradients were measured in the directions of the boule growth and perpendicular to it and the measurements were used to simulate the effect on the specific heat peak. The lower figure shows the actual data compared with the model and the agreement is excellent. Also shown is the expected peak in the absence of a concentration.

measurements of the specific heat peak of a crystal of $Fe_xZn_{1-x}F_2$ with $x = 0.8$ were made with the laser beam oriented along the boule axis direction and perpendicular to it. The effect of the gradient on the data is illustrated in Fig. 3.16. The peak is clearly more rounded for the measurement with the laser beam oriented along the boule axis.

The effect of the gradient on the specific heat critical behavior of $Fe_{0.8}Zn_{0.2}F_2$ has been accurately simulated using the measured (Belanger *et al.*, 1988a; King *et al.*, 1988) variations of $\Delta(n)$ at room temperature. The simulations use the measured variations of T_N with the concentration x and total change in Δn. Figure 3.16 shows the excellent agreement with the specific heat measurements. Also shown is the specific heat peak that would be observed with no variation of the concentration. Clearly, the concentration gradients distort the peak and the critical behavior can be mischaracterized if the gradients are not properly taken into account. Note that the peak appears shifted to lower temperature by the gradients. This is a result of the asymmetry in the specific heat peak. In the absence of a concentration gradient, the data just below the transition temperature are much higher than the data above. That causes the peak to move to a lower temperature when averaging over the variation of concentration. In contrast, when the specific heat is symmetric above and below the transition, the effect of the variation of concentration across the sample is to round the transition, but not to shift the peak.

Figure 3.17 shows what the rounding of the specific heat looks like on a semi-log plot of specific heat. This can be useful when qualitatively

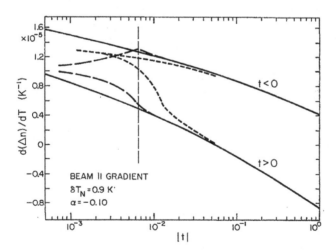

Fig. 3.17: A semi-log plot of the simulation of the effects of the concentration gradients on the specific heat peak (Belanger *et al.*, 1988a). The short-dashed curves are plotted using the peak for the valued of T_N, whereas the long-dashed curves are plotted using the actual T_N. Note that incorrectly using the peak as T_N causes a distortion over a much wider range of $|t|$.

evaluating real data. Effects that show up at small $|t|$ can be identified and attributed to concentration gradients or temperature gradients. Those distorted regions can then be excluded from data fits. Note that using the transition temperature determined from fits to the data rather than the peak temperature results in less distortion of the data away from the transition temperature. Using the fitted transition temperature yields better results in fits to determine the critical behavior.

The effects of gradients on neutron scattering are more difficult to characterize because the critical behavior is obtained from line shape analyses after correcting for the instrumental resolution. The line shapes for the random-exchange model system $Fe_xZn_{1-x}F_2$ are not known beyond mean-field approximations. Also, because the entire sample is used, the effects of gradients can vary in a more complicated manner. Nevertheless, the effects can be approximately simulated. As an example, we can use measurements of $\kappa = 1/\xi$ that would be expected to approach zero, within the limits of the experimental resolution, as T_N is approached. Figure 3.18 a shows κ versus T near the rounded phase transition observed with $x = 0.5$ (Birgeneau et al., 1983a). The concentration variation in this crystal was measured to be $\Delta x \approx 1.3 \times 10^{-2}$ and is not linear. Also shown are the results of two different linear gradient approximations. The crystal seems to have an average gradient between these values, but the curves do not follow the corresponding measured κ versus T behavior precisely. As an example of a crystal with a much smaller gradient, Fig. 3.18 shows κ for a crystal with $x = 0.46$ with a variation of concentration of only $\delta x \approx 2 \times 10^{-4}$. The crystal was cut from an extraordinarily uniform part of the boule, as shown in Fig. 3.19. As expected, κ reaches a much smaller value near the instrumental resolution. For an analysis that goes beyond mean-field line shapes in dilute systems, as described in the following chapter, high quality crystals are essential.

When a transition takes place in a magnetic system with quenched disorder, not only do thermal fluctuations compete with ordering, so do fluctuations of the magnetic interactions from the vacancies. This gives rise in Ising antiferromagnets to the random-exchange (for $H = 0$) and random-field (for $H > 0$) Ising universality classes. The microscopic randomness of magnetic bonds will be uniform if there is no tendency for chemical clustering or anticlustering of like ions on length scales larger than the fluctuation correlation length. This appears to be the case in many of the most studied systems. However, failure to heed the effects of concentration gradients on a macroscopic scale led to controversies in the interpretation of

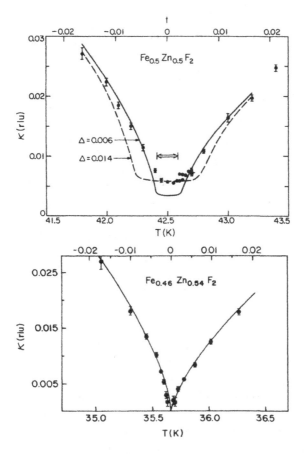

Fig. 3.18: Effect of concentration gradients (Belanger *et al.*, 1988a) on the neutron scattering determinations of κ. The upper figure shows measurements of κ using a crystal with with a variation $\Delta x \approx 1.3 \times 10^{-2}$ along with simulated behaviors with two different linear concentration variations. The lower figure shows similar experimental results using a crystal with a measured variation $\delta x \approx 2 \times 10^{-4}$.

some random-field Ising critical behavior experiments. As will be discussed, random-exchange specific heat peaks, observed with no applied magnetic field in $d = 3$ systems such as $Fe_x Zn_{1-x} F_2$ or $Mn_x Zn_{1-x} F_2$ are asymmetric because $\alpha \neq 0$ and $A^+/A^- \neq 1$. On the other hand, when a field is applied, the random-field Ising specific heat peak becomes nearly symmetric and occurs at a lower temperature. From the previous discussions, it is clear that the $H = 0$ peak will appear shifted downward in temperature by the concentration gradient in the sample, while peak temperature of the

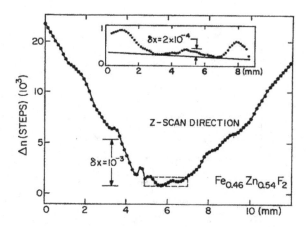

Fig. 3.19: The measured variation of Δn along the growth direction of the $Fe_{0.46}Zn_{0.54}F_2$ boule (King *et al.*, 1988). Using the x dependence of Δn shown in Fig. 3.15, the variation in Δn is converted to a variation in x. The fortuitous minimum near the 6-mm point allowed a reasonably sized sample with an extraordinarily small variation in x to be cut from the boule.

$H \neq 0$ data will not. This effect distorts the shift in temperature from the application of the field. On the other hand, if the shift in the transition temperature is determined by fitting the data, excluding regions that are significantly affected by the variation in concentration, the actual shift in temperature from the field will be correctly determined.

An important question addressed in random-field Ising studies is how the transition temperature changes with the strength of the applied field, which is governed by the random-field crossover exponent, ϕ_{rf}. Some early experiments, for example, (Belanger *et al.*, 1983a), indicated $\phi_{rf} \approx 1.4$, which disagreed with the theoretical prediction (Fishman and Aharony, 1979) that suggested $\phi_{rf} \approx 1.25$. However, it was shown (Aharony, 1986; Calabrese *et al.*, 2003) later that, for site-dilute antiferromagnets in a uniform field, the exponent should be closer to 1.4, as observed in the experiments. The lower exponent value is valid for uniform magnetic systems with a random field applied. For $d = 3$, the generally accepted value for ϕ_{rf} is 1.42(3) for the strongly anisotropic $Fe_xZn_{1-x}F_2$ system, obtained from an extensive number of experiments, and similar values of 1.43(3) and 1.41(5) were obtained for $Mn_xZn_{1-x}F_2$ (Ramos *et al.*, 1988a) and $Fe_xMg_{1-x}Cl_2$ (Leitão and Kleemann, 1987).

Some experimentalists reported that $\phi_{rf} \approx 1.25$, agreeing with the earlier theory. A source of experimental disagreement over the correct value had to do with the effects of the concentration gradients. If the transition temperatures is extracted not from fits to the critical behavior but from the peak temperatures, the lower value is obtained because the $H = 0$ transition is determined to be artificially low, as explained earlier. Eventually, a consensus formed from experimental and theoretical work that the larger value of $\phi_{\rm rf} \approx 1.4$ is correct.

Chapter 4

Critical Behavior Experiments on Anisotropic and Isotropic Antiferromagnets

This chapter covers experiments to study phase transitions in excellent examples of isotropic and anisotropic antiferromagnetic crystals. The $H-T$ phase diagrams will be discussed in pure $d = 3$ systems, followed by the same systems with magnetic site dilution. Next, the critical behavior of pure $d = 2$ and $d = 3$ anisotropic systems will be reviewed. The most conclusive critical behavior investigations have been done on the pure systems and the excellent agreement between experimental results, theory, and computer simulations will be shown. Magnetically diluting these systems provides a straightforward way to study the influences of random-exchange interactions on magnetic phase transitions. Experimental results for $d = 2$ and $d = 3$ dilute anisotropic antiferromagnets will be discussed as well as the excellent agreement of the experimentally observed behaviors with theory and simulation work. For the most part, these experiments are done using the same magnetic systems that were studied for the pure cases, but with a percentage of the magnetic ions randomly replaced by diamagnetic ions, which serve as magnetic vacancies. Using these systems allows a more accurate and informed evaluation of the effects of the magnetic dilution. The study of these same magnetically diluted antiferromagnetic systems with external uniform fields applied is perhaps the most straightforward way to study the effects of random fields on Ising magnetic transitions. An important aspect of random fields as realized in dilute antiferromagnets is that the strength of the random field is proportional to the applied field and can be varied or turned off. The behavior generated by the randomness and frustration in the random-field system is governed by a deceptively simple

Hamiltonian, but this area remains a topic of experimental, theoretical, and computer simulation work. The characterization of the $d = 3$ random-field Ising model remains incomplete, despite much progress and exciting research remains to be done. One particular area where experiments, theory and computer simulations could interact fruitfully to make progress is in the development of more accurate models for the neutron scattering line shapes that could yield better characterizations of some critical behaviors. Finally, experiments near the phase transition in nearly ideal pure, isotropic antiferromagnets will be discussed.

4.1 Crystals

4.1.1 $d = 3$ *Anisotropic crystals*

The most studied systems for experiments on $d = 3$ anisotropic antiferro-magnets are the fluorides listed in Table 4.1. They are the crystals used in the majority of the experiments described in this section. It is useful to list these properties because their different critical behaviors are described and compared. High quality crystals can be grown that are chemical mixtures of these to create mixed magnetic systems or magnetically dilute systems. In addition to those listed, a few of the studies referenced in this chapter used pure and magnetically dilute CoF_2 and $FeCl_2$.

4.1.1.1 *Crystal structure of FeF_2, MnF_2, and ZnF_2 and magnetic interactions of FeF_2 and MnF_2*

The crystal structure of FeF_2, MnF_2, ZnF_2, and their mixtures, is shown in Fig. 4.1. The metal ions form a body-centered tetragonal lattice with lattice parameters $a = b$ and c shown in the figure and listed in Table 4.1. The spins align with the c axis. When ordered, the corner sites and body-centered sites form two sublattices that order antiferromagnetically with respect to each other. The fluorine ions are at relative positions u and $1-u$ along the diagonals of the metal ions on each sublattice. The magnetic Hamiltonian of the pure systems with no magnetic field applied is

$$H = \sum_{i,j} J_{i,j} \vec{S}_i \cdot \vec{S}_j - \sum_i D(S_i^z)^2. \tag{4.1}$$

The magnetic interactions between the metal ions are superexchange interactions involving the F ions. The fluorine ions of the two sublattices are rotated $90°$ with respect to each other. The two sublattices are equivalent

Table 4.1: Crystal parameters.

	FeF_2	MnF_2	ZnF_2
a (Å)	4.70	4.87	4.70
c (Å)	3.31	3.31	3.13
u	0.300	0.305	0.303
J_1 (K) $(z_1 = 2)$	−0.04	−0.32	—
T_1 (K)	0.3	3.7	—
J_2 (K) $(z_2 = 8)$	2.62	1.76	—
T_2 (K)	83.8	82.1	—
J_3 (K) $(z_3 = 4)$	0.27	0.04	—
T_3 (K)	−4.3	0.27	—
D (K)	9.29	0.27	—
S	2	5/2	—

Notes: The rutile structure is shown in Fig. 4.1. The Hamiltonian is $H = \sum_{i,j} J_{i,j} \vec{S}_i \cdot \vec{S}_j - \sum_i D(S_i^z)^2$. The dominant interaction is J_2 between the corner and center ions. The effective temperature for each interaction is given by $T_m = (-1)^m (2/3) z_m J_m S(S + 1)$. The corner ions constitute one sublattice and the center ions constitute the other sublattice; they are equivalent except that the fluorine ions are rotated by a relative angle of 90°. The exchange and anisotropy parameters for FeF_2 are from Hutchings *et al.* (1970). The magnetic percolation threshold concentration, considering only the J_2 interactions and random dilutions on the local scale, is site dilution value $x = 0.246$ (Stauffer and Aharony, 1994). The exchange parameters for MnF_2 are from Nikotin *et al.* (1969). The anisotropy parameter for MnF_2 is from Barak *et al.* (1978). ZnF_2 is diamagnetic and serves to effectively introduce magnetic vacancies. The lattice parameters are from Baur (1958) and Stout and Reed (1954).

unless the lattice is distorted along the a–b diagonal, in which case the fluorines are affected differently. ZnF_2 is diamagnetic, but for the purposes of this book, the substitution of Mn or Fe ions with Zn represents the creation of a magnetic vacancy.

The systems above grow well as mixed crystals, such as the magnetically diluted systems $Fe_x Zn_{1-x} F_2$ and $Mn_x Zn_{1-x} F_2$ and the mixed anisotropy strength system $Fe_x Mn_{1-x} F_2$. It is not always clear how the concentration is determined in experimental reports with mixed systems. One method that works well to determine x with an accuracy of about 1% is to use Archimedes principle if it can be used without damage to the samples. Once the transition temperature T_N versus x for a particular system

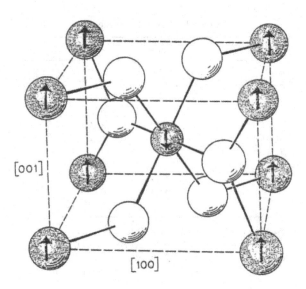

Fig. 4.1: FeF_2 structure. The shaded circles represent the Fe ions and the light-colored circles are the F ions. The magnetic ions form a tetragonal lattice with superexchange interactions, represented by the solid bonds, between the corner and center ions via the fluorine ions. When ordered, the body-center spins point along the unique c axis and the corner spins order in the opposite direction, forming the two antiferromagnetic sublattices.

has been measured, the concentration can be determined from the transition temperature. Section 3.7 describes the importance of macroscopic uniformity of the constituents throughout the crystal used in the experiments and the importance of characterizing the uniformity in order to interpret experimental data correctly. Another question relevant to interpretation is the uniform randomness of the constituents on the atomic scale. For the systems covered in this and the next chapter, there is no evidence of one ion clustering or anticlustering with like ions on the atomic scale. It is difficult to ascertain the local randomness, but there is evidence in the behaviors around percolation threshold concentrations. As will be described, behaviors near the magnetic and vacancy percolation threshold concentrations correlate well with the concentrations calculated for body-centered tetragonal (the same as body-centered cubic) lattices. An indication that constituents are randomly positioned is the scattering from the fractal structures that occur very close to the calculated vacancy percolation threshold concentration (Barber *et al.*, 2004). A small amount of local nonrandomness will not affect experiments if the constituents are not correlated on length scales comparable to the fluctuation correlations that are being studied.

4.1.1.2 Crystal structure of FeCl$_2$

Another $d = 3$ system that has played a role in the study of the random-exchange and random-field Ising models is the rhombohedral structured FeCl$_2$ and magnetically diluted Fe$_x$Mg$_{1-x}$Cl$_2$. The magnetic structure consists of ferromagnetically ordered layers that form antiferromagnetically stacked layers. A first-order transition is observed in a field in FeCl$_2$ (Kushauer and Kleemann, 1995) so the phase diagram exhibits tricritical behavior that is distinct from the FeF$_2$ and MnF$_2$ systems described above which exhibit bicritical points (King and Rohrer, 1979).

4.1.2 d = 2 Anisotropic crystals

The isomorphic crystals Rb$_2$CoF$_4$ and K$_2$CoF$_4$ have proven to be excellent realizations of the $d = 2$ Ising universality class. The crystal structure is the same as that of K$_2$NiF$_4$ shown in Fig. 4.2. The magnetic properties are characterized by Breed *et al.* (1969), Ikeda and Hutchings (1978), and Lines (1967). The planes of the magnetic Co ions on a square lattice are well separated. Because the superexchange interaction falls off rapidly with distance between the Co ions, the interaction between the planes is several orders of magnitude smaller than that within the planes. The resulting order at the antiferromagnetic transition temperature, T_N, is well-characterized by the $d = 2$ Ising universality class, though $d = 3$ ordering slowly builds with time. The fluctuation correlation length, ξ, primarily grows within the planes, so the two-dimensional scattering manifests itself as scattering rods in reciprocal space, while the $d = 3$ scattering is at antiferromagnetic Bragg points. This separation of the nuclear and antiferromagnetic scattering allows scattering experiments that probe only the $d = 2$ magnetic ordering processes.

4.1.3 d = 3 Isotropic crystals

The cubic structures of KNiF$_3$ and RbMnF$_3$ have the KNiF$_3$ structure shown in Fig. 4.2. The magnetic properties are described in Lines (1967) and Teaney *et al.* (1962). As described in Section 4.2, the anisotropies are extremely small, making them ideal realizations of the $d = 3$ Heisenberg universality class.

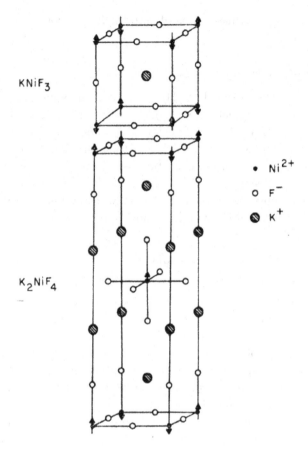

KNiF$_3$

K$_2$NiF$_4$

• Ni^{2+}

○ F$^-$

◉ K$^+$

Fig. 4.2: KNiF$_3$ and K$_2$NiF$_4$ structure (Lines, 1967). Reproduced with permission from American Physical Society.

4.2 Phase Diagram Measurements of Anisotropic and Isotropic Antiferromagnets

4.2.1 *The bicritical and tricritical points in pure and dilute anisotropic antiferromagnets*

Figure 4.3 shows a typical $H_\parallel$ versus T phase diagram for a $d = 3$ anisotropic antiferromagnet. $H_\parallel$ is the field aligned precisely along the unique easy axis. A bicritical point occurs at $H_\parallel = H_B$ and $T = T_B$. As T is reduced with $H < H_B$, a second-order transition occurs from the paramagnetic state to an antiferromagnetic long-range ordered state. The $d = 3$ Ising universality class ($n = 1$) along this boundary is the same as it is for $H = 0$. For a range

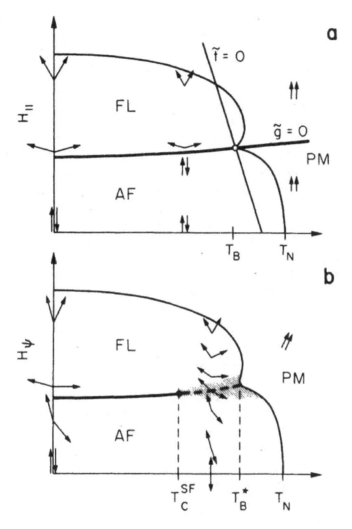

Fig. 4.3: The $d = 3$ $H_\parallel$ versus T phase diagram of an anisotropic antiferromagnet showing the bicritical point when the field is carefully aligned with the easy axis of the crystal (upper panel) and the behavior when the field is slightly misaligned (lower) (King and Rohrer, 1979) For the carefully aligned case, the boundary between the paramagnetic (PM) phase and antiferromagnetic (AF) phase at low fields is second-order with critical behavior belonging to the Ising universality class. The boundary between the PM and spin-flop (SF) phase is second-order with XY universality. The line $\tilde{g} = 0$ is tangential to the the first-order phase boundary between the AF and SF ordered phases at the bicritical point. All three phase boundaries meet at the bicritical point, at which there is a second-order phase transition of the Heisenberg universality class. When the crystal is slightly misaligned, the bicritical point is not properly probed. Reproduced with permission from American Physical Society.

of field $H > H_B$, a second-order transition takes place from paramagnetic disorder to a spin-flop phase, where the two sublattices have equal antiparallel ordered components perpendicular to $H_\parallel$ and equal components along $H_\parallel$. The ordering of the perpendicular components belongs to the $d = 3$ XY universality class ($n = 2$). When $H_\parallel$ is sufficiently large, no transition takes place and the paramagnetic phase extends to $T = 0$.

A first-order transition boundary extends below T_B between the spin-flop and antiferromagnetic phases. The line $\tilde{g} = 0$ is tangent to the first-order boundary at the bicritical point. The axes perpendicular and parallel to it are used to characterize the curvature of the second-order transition boundaries. Using the linear scaling fields $\tilde{g} = g - pt$ and $\tilde{t} = t + qg$, where $t = (T - T_B)/T_B$ and $g = H_\parallel^2 - H_B^2$, the curvature is given by $\tilde{g} = \pm w_\pm \tilde{t}^\phi$, where ϕ is the crossover exponent and the amplitudes w_+ and w_- are for the spin-flop and antiferromagnetic boundaries, respectively. As the bicritical point is approached along the $\tilde{g} = 0$ line, the critical behavior is that of the $d = 3$ Heisenberg universality class ($n = 3$).

The values of T_B and H_B are determined by the exchange and anisotropy fields, H_E and H_A, respectively, of the antiferromagnet. As $H_A \to 0$, the system is a Heisenberg antiferromagnet and the bicritical point approaches the $H_\parallel = 0$ axis at T_N.

The bicritical point can be reached with reasonable applied magnetic fields in MnF$_2$, so it is an ideal system to study. The phase diagram has been characterized in detail (King and Rohrer, 1979; Shapira and Becerra, 1976). Important features (King and Rohrer, 1979) near the bicritical point at $H_\parallel = 118.353(10)$kOe and $T_B = 64.792(1)$ K are shown in Fig. 4.4. The applied field was carefully aligned with the unique crystalline axis, which allowed accurate determination of the curvature of the phase boundaries. The coordinates described above were used in fits to the data from which the crossover exponent $\phi = 1.279(31)$ was determined. That result is in good agreement with the theoretical value 1.250(15) (Pfeuty *et al.*, 1974).

With $H_E = 554.7$kOe and $H_A = 190.7$kOe, the phase diagram and equilibrium bicritical point of the more anisotropic $d = 3$ antiferromagnet FeF$_2$ is not readily accessible with steady applied fields, but the phase diagram has been studied (Jaccarino *et al.*, 1983) with pulsed magnetic fields along with the magnetically dilute system Fe$_x$Zn$_{1-x}$F$_2$ (King *et al.*, 1983).

The field at which the bicritical point occurs depends on the strength of the anisotropy and should be at $H = 0$ for isotropic antiferromagnets; it is then the same as the Heisenberg transition at T_N. The phase diagrams of

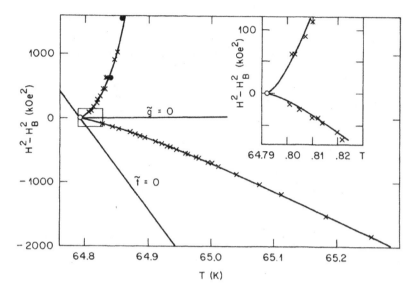

Fig. 4.4: The $d = 3$ $H^2 - H_B^2$ versus T phase diagram of the anisotropic antiferromagnet MnF_2 near the bicritical point with an accurately aligned field (King and Rohrer, 1979). Reproduced with permission from American Physical Society.

the extremely isotropic $d = 3$ antiferromagnets $RbMnF_3$ and $KNiF_3$ were studied near T_N (Shapira and Becerra, 1977; Shapira and Oliveira, 1978). The example of $RbMnF_3$ is shown in Fig. 4.5 and the crossover exponent $\phi = 1.26$ is in good agreement with theory and the bicritical point study in MnF_2.

For $d = 2$, the bicritical point of Rb_2MnF_4 has been studied using neutron scattering (Cowley *et al.*, 1993) and is shown in Fig. 4.6. A simple expectation might be that the bicritical point would move to $T = 0$ since a $d = 2$ Heisenberg magnet should not order. However, the data in Fig. 4.6 seem to show it to be at finite T. The authors attribute this possibly to the interactions between magnetic planes and to cubic anisotropy. A slight misalignment of the field could also contribute.

Whereas the $H - T$ phase diagrams of FeF_2 and MnF_2 show a bicritical point, the metamagnet $FeCl_2$ in a field has a tricritical point, with $T_t = 20.5$ K and $T_N = 23.0$ K (Shang and Salamon, 1980). The $H-T$ phase diagram (Birgeneau *et al.*, 1974) of $FeCl_2$ is shown in Fig. 4.7. Instead of a bicritical point, the second-order phase boundary separating the paramagnetic and antiferromagnetic states at low temperature ends at a tricritical point. This is more clearly seen in the $M - T$ phase diagram of Fig. 4.8,

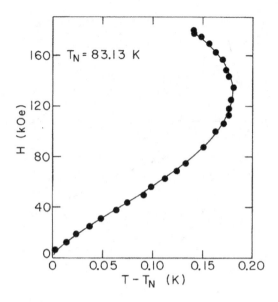

Fig. 4.5: The phase diagram of isotropic $d = 3$ RbMnF$_3$ (Shapira and Becerra, 1977). The alignment of the field is not critical for an isotropic case. Reproduced with permission from American Physical Society.

which shows the mixed phase between the paramagnetic phase and antiferromagnetic phase for $T < T_t$. The critical behavior near a tricritical point can exhibit mean-field exponents.

While the $d = 3$ and $d = 2$ phase diagrams are well characterized in the pure systems, magnetic dilution complicates the phase diagrams. Metastability plays an essential role, as will be seen in the discussion of the random-field Ising model in Sections 4.7.1 and 4.7.2. In addition to metastability issues, other complexities of the phase diagram of the site-dilute anisotropic antiferromagnet have been theoretically investigated (Aharony, 1978a,b). With small random fields induced by the applied field along the uniaxial direction, the phase diagrams are similar to those in pure systems with reduced dimensionality. As with pure systems, the spin-flop and antiferromagnetic regions can still be separated by a first-order transition boundary, and a mixed phase region can exist near the multicritical point, which can be either bicritical or tetracritical.

A phase diagram for Mn$_{0.75}$Zn$_{0.25}$F$_2$ based on dilatometry, ultrasonic-attenuation measurements, and magnetometry (Shapira *et al.*, 1984) strikingly contrasts in shape that of pure MnF$_2$, as shown in Fig. 4.9. The basic

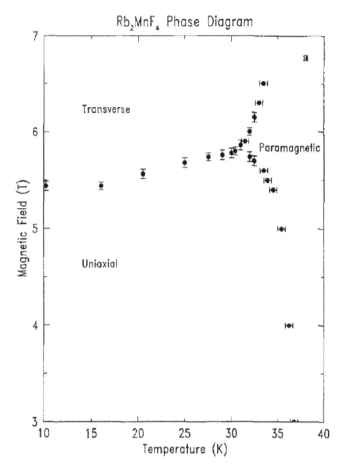

Fig. 4.6: The phase diagram of anisotropic $d = 2$ Rb$_2$MnF$_4$ measured using neutron scattering (Cowley *et al.*, 1993). Reproduced with permission from Springer.

phase boundaries do appear, however. The phase diagrams for $d = 3$ Mn$_x$Zn$_{1-x}$F$_2$ for $x = 0.75$ and $x = 0.5$, studied using neutron scattering (Fig. 4.10) show the complications, at least for $x = 0.75$, introduced by metastability associated with the random-fields generated by the applied field (Cowley *et al.*, 1989). Between the metastability line and the long-range order boundary, possible mixed phases were observed. The behavior near the multicritical point with a smaller amount of dilution was explored in detail with neutron scattering in another dilute antiferromagnet, Rb$_2$Fe$_{0.92}$In$_{0.08}$Cl$_5$·H$_2$O (Palacio *et al.*, 1998). As depicted in Fig. 4.11, in addition to the primary phase boundary between the antiferromagnetic and

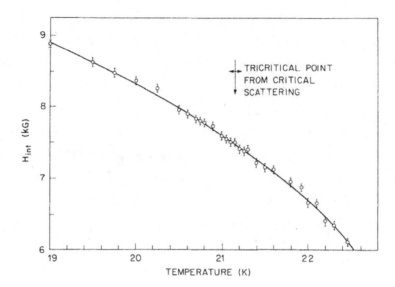

Fig. 4.7: The H–T phase diagram of FeCl$_2$ (Birgeneau *et al.*, 1974). The bicritical point is indicated. The phase boundary at higher temperatures is second-order and at lower temperatures first order. Reproduced with permission from American Physical Society.

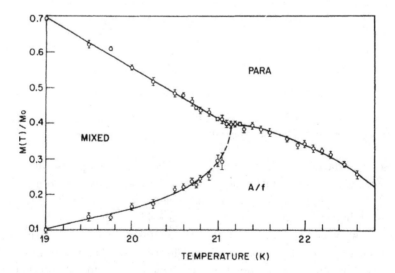

Fig. 4.8: The M–T phase diagram of FeCl$_2$ (Birgeneau *et al.*, 1974). The first-order character for temperatures below the tricritical point is clear in this phase diagram from the region with mixed phases below the tricritical point. Reproduced with permission from American Physical Society.

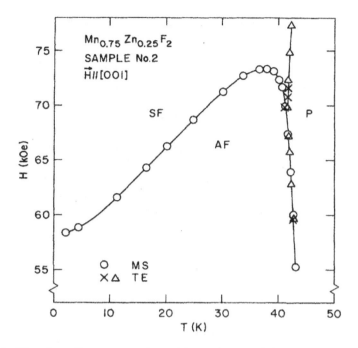

Fig. 4.9: The phase diagram near the multicritical point of the diluted antiferromagnet $Mn_{0.75}Zn_{0.25}F_2$ (Shapira *et al.*, 1984). Reproduced with permission from American Physical Society.

spin-flop boundaries, secondary scattering peaks were observed at slightly higher fields; this is particularly evident in the inset of the figure. It is suggested that this regions possibly represents the theoretically predicted intermediate phase or that the inhomogeneities prevent the spin-flop state from spreading uniformly throughout the sample.

While the effects from metastability are not as clear for $x = 0.5$ in Fig. 4.10 in the neutron scattering experiments, magnetometry measurements (Montenegro *et al.*, 1992) clearly demonstrate their influence in the phase diagram, as shown in the upper panel of Fig. 4.12.

The phase diagram of $Mn_{0.39}Zn_{0.61}F_2$ (Rosales-Rivera *et al.*, 2001), shown in the lower panel of Fig. 4.12 appears very similar to the higher concentrations but with smaller fields and temperatures. Interestingly, for a slightly lower concentration, $x = 0.35$, the phase diagram exhibits spin-glass-like behavior for all fields (Montenegro *et al.*, 1995), much like the behavior of $Fe_xZn_{1-x}F_2$ for x near the magnetic percolation threshold $x_p = 0.25$. The spin-glass-like behavior will be discussed in Chapter 5.

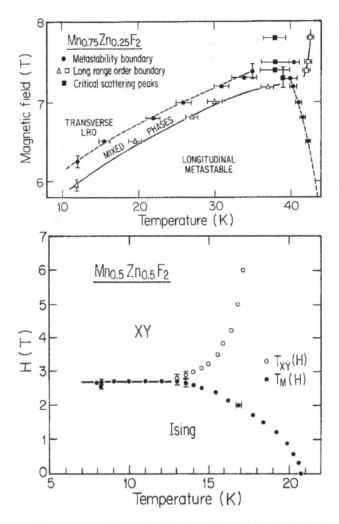

Fig. 4.10: The phase diagram near the multicritical point of the diluted antiferromagnet $Mn_{0.75}Zn_{0.75}F_2$ and $Mn_{0.5}Zn_{0.5}F_2$ (Cowley *et al.*, 1989). Reproduced with permission from Springer Nature.

$Mn_{0.4}Zn_{0.6}F_2$ was studied using birefringence (Ramos *et al.*, 1988a). As soon as a field is applied, the lower part of the diagram is governed by random fields. The upper boundary is not observed in birefringence measurements, as shown in Fig. 4.13. However, the point where the boundary stops scaling with $H^{2/\phi_{rf}}$ corresponds roughly to the intersection of the boundaries in Fig. 4.12 for $x = 0.39$.

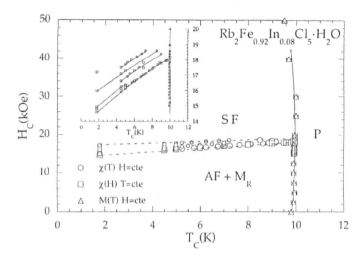

Fig. 4.11: The phase diagram near the multicritical point of the diluted antiferromagnet $Rb_2Fe_{0.92}In_{0.08}Cl_5 \cdot H_2O$ (Palacio *et al.*, 1998).

The magnetically dilute $d = 2$ Ising case $Rb_2Mn_{0.7}Mg_{0.3}F_4$ was also investigated and has the additional complexity of random fields that destroy the Ising transition, as described in Section 4.7.1. It appears that the upper transverse ordering transition boundary is headed to meet the lower transition boundary at $T = 0$, but intersects the metastability boundary just above $T = 0$, as shown in Fig. 4.14.

4.2.2 The influence of anisotropy on the transition T_N

Small fields applied to a nearly isotropic antiferromagnet increase the transition temperature rapidly, as we see from the phase diagram in Fig. 4.5. The same effect occurs near the bicritical point, as seen in Fig. 4.3; increasing or decreasing the field rapidly increases the transition temperature as XY or Ising anisotropy is introduced. The effect on the transition temperature can also be observed in zero field by changing the intrinsic Ising anisotropy in a mixed anisotropy system.

The increase in the transition temperature T_N at $H = 0$ with increasing anisotropy was studied (Belanger *et al.*, 1982a) in the mixed magnetic system $Fe_xMn_{1-x}F_2$. Although this system is magnetically dense, the Fe ions experience mostly a large single-ion anisotropy from the lattice and the Mn ions see a much smaller dipolar anisotropy. So not only is the anisotropy strength random from site to site, the average anisotropy

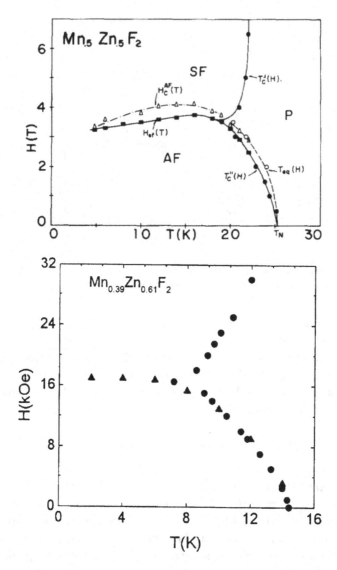

Fig. 4.12: The phase diagram near the multicritical point of the diluted antiferromagnet Mn$_{0.5}$Zn$_{0.5}$F$_2$ (Montenegro *et al.*, 1992) and Mn$_{0.39}$Zn$_{0.61}$F$_2$ (Rosales-Rivera *et al.*, 2001). Reproduced with permission from Elsevier.

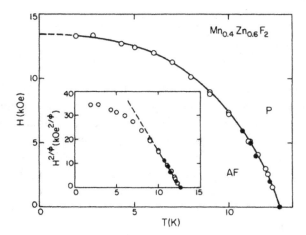

Fig. 4.13: The phase diagram of $Mn_{0.4}Zn_{0.6}F_2$ (Ramos *et al.*, 1988a). The solid symbols were determined from $d(\Delta n)/dT$ and the open symbols are from $d(\Delta n)/dH$. The straight line in the inset shows where the data start to deviate from the $H^{2/\phi_{rf}}$ behavior. Reproduced with permission from American Physical Society.

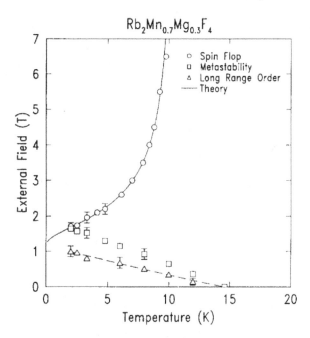

Fig. 4.14: The phase diagram of $Rb_2Mn_{0.7}Mg_{0.3}F_4$ (Birgeneau *et al.*, 1991). Reproduced with permission from Elsevier.

increases with x. The mean-field prediction for the transition temperature in the zero anisotropy case is given by

$$T_N(0) = 2/3zJS(S+1), \qquad (4.2)$$

where $z = 8$ is the body-centered coordination number, S is the spin, and J is the effective exchange interaction. For the mixed system, an approximate calculation can be made for the x dependence of T_N assuming a weighting of T_N from Eq. (4.2) corresponding to the concentration of Mn and Fe ions. Because the two values for FeF_2 and MnF_2, $T_N(0) = 61.0$ and 61.7 K, respectively, are nearly identical, the change in T_N with x from the mean-field calculation is small. The observation of the rapidly changing transition temperature seen in Fig. 4.15 can then be attributed largely to the average strength of the anisotropy for small values of x. The anisotropy

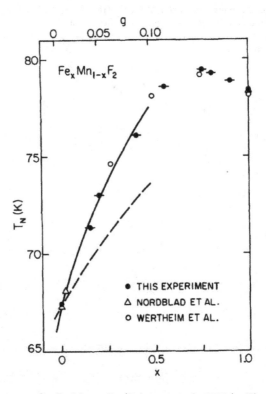

Fig. 4.15: T_N versus x for $Fe_x Mn_{1-x} F_2$ (Belanger *et al.*, 1982a). The steep rise in the transition temperature near $x = 0$ is a consequence of the increase of the anisotropy, which is much larger in FeF_2 than in MnF_2. The dashed line is a mean-field prediction, whereas the solid line is a fit to a scaling prediction (Pfeuty *et al.*, 1974).

dependence for small x is predicted (Pfeuty *et al.*, 1974) to scale as

$$T_N(g) = T_N(0)(1 + Ag^{1/\phi}), \tag{4.3}$$

where g represents the strength of the anisotropy, A is of order unity, and ϕ is the crossover exponent. The anisotropy is approximated by

$$g = \frac{D}{zJ}\frac{(2S-1)(2S+3)}{10S(S+1)}, \tag{4.4}$$

using measured values (Section 4.1) of the anisotropy term D for MnF_2 and FeF_2 and weighting by the concentration of Mn and Fe ions. Figure 4.15 shows a fit to this behavior for $A = 1.1$, which is quite good for the low anisotropy region. Note that $g = 0$ is below $x = 0$ because MnF_2 has a small amount of anisotropy. At larger values of x, the behavior is dominated by the interaction strength between the Fe and Mn ions, which is likely slightly stronger than the interaction strength between Fe ions or between Mn ions.

Because $dT_N(x)/dT = 0$ near $x = 0.75$, a concentration gradient in this case should not round the transition. Figure 4.16 demonstrates that this

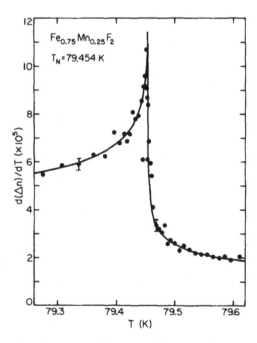

Fig. 4.16: The magnetic specific heat versus T, measured for $Fe_{0.75}Mn_{0.25}F_2$ (Belanger *et al.*, 1982a). The peak sharpness and shape are nearly identical to that of FeF_2.

is the case, with a peak as sharp as that observed for FeF_2. The shape of the curve is identical to that of FeF_2, showing that there is no discernible crossover to random-exchange behavior since the exchange interaction is nearly the same between the different magnetic ions.

4.3 Specific Heat Critical Behavior of Pure Ising Antiferromagnets

4.3.1 $d = 2$ *Pure Ising specific heat critical behavior*

The isomorphs Rb_2CoF_4 and K_2CoF_4 are nearly ideal $d = 2$ Ising antiferromagnets and this provides excellent opportunities to explore the critical behavior of this universality class. The symmetric, logarithmic divergence of the specific heat in the $d = 2$ Ising model, predicted by the exact Onsager solution (Onsager, 1944), is apparent in $d(\Delta n)/dT$ versus T measurements using Rb_2CoF_4 (Nordblad *et al.*, 1983b) shown in Fig. 4.17. The behaviors of the specific heat versus $|t|$ for $T < T_N$ and $T > T_N$ are the same over two decades of reduced temperature and represent a straight-line behavior in the semi-log plot indicating a symmetric, logarithmic divergence, as predicted. Fits to the critical behavior using

$$d(\Delta n)/dt = (A/\alpha)(|t|^{-\alpha} - 1) + B + Et, \tag{4.5}$$

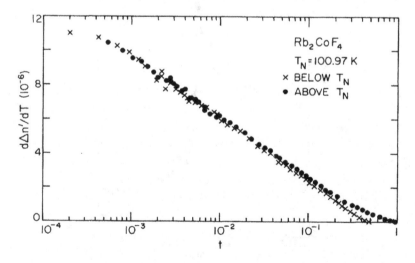

Fig. 4.17: The logarithmic specific heat critical behavior of Rb_2CoF_4 (Nordblad *et al.*, 1983b). Reproduced with permission from American Physical Society.

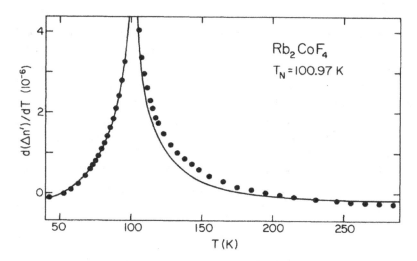

Fig. 4.18: The specific heat of Rb_2CoF_4, represented by $d\Delta n/dT$ versus T, compared to the exact $d = 2$ Ising model solution. A small constant background was subtracted from the exact solution to account for the nonmagnetic contributions to the data (Nordblad *et al.*, 1983b).

yield the fitted values $\alpha = -0.002(13)$ and $A^+/A^- = 1.010(8)$ over the range $5 \times 10^{-4} < |t| < 5 \times 10^{-2}$, which is consistent with the exact solution. The agreement with the exact $d = 2$ Ising prediction holds remarkably well even over a large range in temperature, well beyond the asymptotic symmetric, logarithmic divergence, as shown in Fig. 4.18. The small deviations between experiment and theory well above T_N are a result of the system being an anisotropic antiferromagnet and not a perfect Ising system. The higher energy levels of the $S = 2$ moments can be excited at elevated temperatures. The accuracy of the measurements relies on the relative insensitivity of the birefringence technique to the phonon contributions to the specific heat, as explained in Section 3.2. Earlier attempts to characterize the critical behavior using conventional specific heat were hampered by the large phonon background (Ikeda *et al.*, 1976).

4.3.2 $d = 3$ *Pure Ising specific heat critical behavior*

For the $d = 3$ Ising universality class, there are no exact calculations of the critical parameters but, for the pure case, theory and simulations provide accurate predictions, as listed in Tables 2.1 and 2.2 that can be tested in high precision experiments. In direct measurements of the specific heat versus T in anisotropic $d = 3$ systems in zero applied magnetic field, extraordinarily accurate critical behavior measurements were obtained for FeF_2 (Chirwa *et al.*, 1980) and MnF_2 (Nordblad *et al.*, 1981). The large phonon contributions were subtracted using previously measured (Boo and Stout, 1976; Stout and Catalano, 1955) background data for FeF_2 and MnF_2. Data at reduced temperatures as small as $|t| = 7 \times 10^{-5}$ were obtained, allowing access to the $d = 3$ asymptotic Ising critical region. The birefringence technique ($d(\Delta n)/dT$ versus T) achieved similarly accurate measurements (Belanger *et al.*, 1983b). The small phonon background contributions were subtracted by fitting to Einstein and Debye functions. It was shown (Belanger, 1981) that the lattice contributions in diamagnetic crystals ZnF_2 and MgF_2 could be fit in this way. The magnetic specific heat of MnF_2 was accurately determined up to $T = 550$ K in this manner, where $T_N = 67$ K.

Semi-log plots of the specific heat and birefringence data, with the appropriate nonmagnetic contributions subtracted, are compared in Figs. 4.19–4.22. The asymmetry of the specific heat peak is evident from the smaller specific heat amplitude above T_N and the asymptotic divergence is indicated by a positive value for α. The data for the more anisotropic FeF_2 exhibit positive curvature, consistent with a positive critical exponent α, over a wider range, $|t| < 10^{-2}$, than the positive curvature range, $|t| < 10^{-3}$, for MnF_2. This is a result of the stronger effects of the crossover

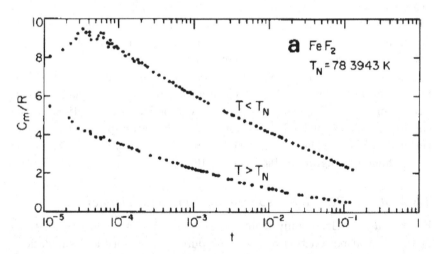

Fig. 4.19: The magnetic specific heat critical behavior of FeF_2 (Chirwa *et al.*, 1980). Reproduced with permission from Elsevier.

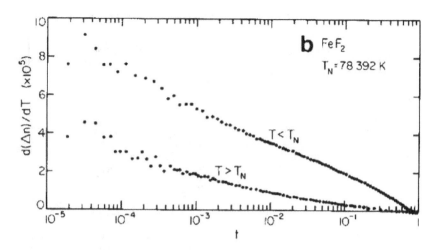

Fig. 4.20: The magnetic specific heat critical behavior of FeF_2 measured using $d(\Delta n)/dT$ (Belanger *et al.*, 1983b).

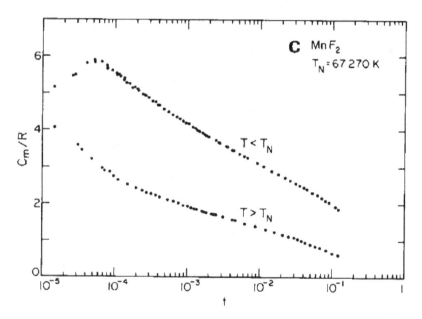

Fig. 4.21: The magnetic specific heat critical behavior of MnF_2 (Nordblad *et al.*, 1981). Reproduced with permission from Elsevier.

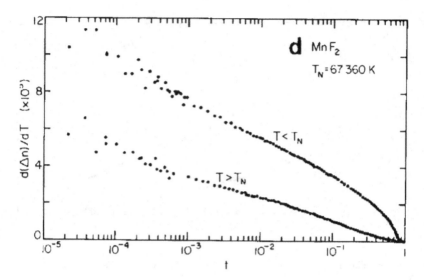

Fig. 4.22: The magnetic specific heat critical behavior of MnF$_2$ measured using $d(\Delta n)/dT$ (Belanger *et al.*, 1983b).

from Ising critical behavior at small $|t|$ towards Heisenberg behavior at larger $|t|$ for the MnF$_2$ system.

The data for both systems were analyzed using

$$C = (A^\pm/\alpha)|t|^{-\alpha}(1 + D^\pm|t|^x) + B + Et, \qquad (4.6)$$

where the asymptotic power is represented by the leading power law, with exponent α and where the subscripts $-$ and $+$ refer to $t < 0$ and $t > 0$, respectively. Within the parentheses is a noncritical crossover correction to the power law behavior. The exponent x is set equal to an approximate value of 0.5 (Ahlers *et al.*, 1974; Kornblit and Ahlers, 1973). Results from fits to data from both experimental techniques, shown in Table 4.2, are consistent, attesting to the accuracy of the nonmagnetic background subtractions. The values obtained for the universal parameters α and A^+/A^- for small $|t|$ and $D = 0$, listed in Table 4.2 are in excellent agreement with theory and Monte Carlo results shown in Tables 2.1 and 2.2. When α is fixed to the value 0.11 and data are fit over a wide range of $|t|$, the values of D^- and D^-/D^+ are obtained. As expected, the values of the crossover effects are greater in MnF$_2$, as demonstrated by the larger value of D^-. The expected range of influence of crossover effects is predicted to be for $|t| > (H_A/H_E)^{1/\phi}$, where H_A and H_E are the effective anisotropy

Table 4.2: $d = 3$ Specific heat critical parameters from fits to Eq. (4.6).

Exponent	FeF$_2$	MnF$_2$
Specific Heat		
with $D^-, D^+ = 0$		
fit range	$7 \times 10^{-5} < |t| < 5 \times 10^{-3}$	$10^{-4} < |t| < 5 \times 10^{-3}$
α	0.111(7)	0.123(5)
A^+/A^-	0.543(20)	0.491(14)
$d(\Delta n)/dT$		
with $D^-, D^+ = 0$		
fit range	$7 \times 10^{-5} < |t| < 5 \times 10^{-3}$	$10^{-4} < |t| < 5 \times 10^{-3}$
α	0.115(4)	0.091(5)
A^+/A^-	0.543(20)	0.596(15)
Crossover		
with $\alpha = 0.110$		
fit range	$10{-}4 < |t| < 5 \times 10^{-2}$	$10{-}4 < |t| < 5 \times 10^{-2}$
D^-	$-0.175(75)$	$-0.65(15)$
D^+/D^-	0.5(8)	1.85(65)

and exchange energy fields, and $\phi = 1.25$ is the crossover exponent (see Chapter 2). The values of $(H_A/H_E)^{1/\phi}$ are 0.035 and 0.41 for FeF$_2$ and MnF$_2$, respectively.

It is clear from the above discussion that the $d = 3$ Ising asymptotic critical behavior of MnF$_2$ cannot be characterized unless data for $|t| < 10^{-3}$ can be evaluated. This is because MnF$_2$ is a weakly anisotropic antiferromagnet. By itself, it is problematic when it is characterized as either being an Ising system or a Heisenberg system as it often is. It is studied in most of the experiments covered in this book for two reasons. First, its bicritical point is readily accessible. Second, it serves as a contrast to the strongly anisotropic FeF$_2$ antiferromagnet. However, systems with strong crossover effects can be problematic to analyze and the specific heat shows this quite well.

Figure 4.23 shows data taken with an ac specific heat technique (Ikeda *et al.*, 1978). Although data were taken for $|t|$ as small as 10^{-4}, there are clear differences with the data in the previous figures. This might have to do with the ac technique. It is difficult to ensure that the sample is truly in equilibrium with that technique. The data analysis for $|t|$ clearly show a curvature consistent with $\alpha < 0$, suggesting Heisenberg critical behavior, but the authors also conclude that the exponent α is different above and below T_N. The anomalous behavior is just a manifestation of Heisenberg to Ising crossover that is only apparent if data with much smaller

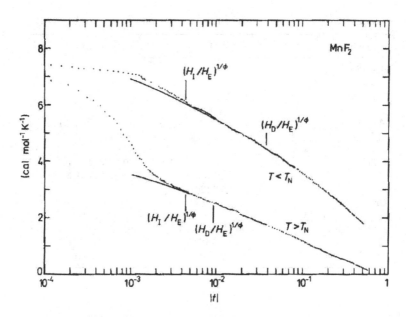

Fig. 4.23: The specific heat of MnF$_2$ measured using an ac specific heat technique (Ikeda *et al.*, 1978). Reproduced with permission from IOP Publishing.

$|t|$ are obtained. None of these issues would be encountered with FeF$_2$. To characterize data for a particular universality class, the possibility of crossover behavior has to be considered. It is not always possible to acquire data extremely close to the transition, so a healthy amount of caution must be part of the data analysis. Specific heat measurements are particularly susceptible to crossover effects.

The experiments for $d = 2$ and $d = 3$ set a standard by which magnetically dilute Ising systems can be evaluated and interpreted. The dilute systems pose challenges, particularly from metastability effects and the rounding due to the inevitable concentration gradients in the samples. The lesson from the experiments on MnF$_2$ is that crossover effects can mask the correct asymptotic universal behavior. Fits to experimental data taken over a smaller temperature range have yielded negative values for α, but that was simply a consequence of fitting data that were not close enough to T_N and did not account for the crossover effects. Particularly with specific heat measurements, effective exponents must always be a concern in characterizing new universality classes and it is important in magnetically dilute systems to do measurements using different concentrations and, when appropriate, different applied fields to assess whether the measured exponents represent the asymptotic behavior or are effective exponents significantly influenced by crossover. Unlike the pure cases, for which the universal critical behavior is well characterized by theory and simulations, the critical behavior of magnetically dilute systems, especially of the specific heat, is not as well known and experiments play a crucial role. This is particularly true of the random-field Ising model.

4.4 The Order Parameter of Pure Ising Antiferromagnets

The degree of long-range ordering in an antiferromagnetic system below the transition temperature is described by the staggered magnetization

$$M_S = M_0 |t|^\beta, \tag{4.7}$$

where β is the order-parameter exponent characteristic of the universality class. Experimentally, the exponent appears to be well characterized at reduced temperatures larger than is possible with the specific heat. Also, when a crossover from one behavior to another as t approaches zero occurs, it apparently does so quickly. This has been observed in the magnetically diluted systems discussed below. For the pure systems, the exponents measured agree well with theory and simulations.

4.4.1 $d = 2$ *Pure Ising order parameter*

For the $d = 2$ Ising antiferromagnets, such as Rb_2CoF_4 and K_2CoF_4, the two-dimensional scattering intensity is along rods in reciprocal space. Scattering at the three-dimensional Bragg scattering involves ordering between the weakly coupled two-dimensional magnetic layers. As we will see for the $d = 3$ case, extinction effects can prevent characterization using neutron scattering. However, the $d = 2$ scattering along the rods between the $d = 3$ Bragg points can be used without significant extinction effects because the scattering is spread out over the rods and so is not intense at any one point in reciprocal space. The beam is therefore not depleted in the way it is for $d = 3$ systems. The line shapes in the direction perpendicular to the rods represent the scattering from magnetic fluctuations for $q \neq 0$ and the square of the order parameter for $q = 0$. Neutron scattering measurements (Hagen and Paul, 1984) of the order parameter are shown in Fig. 4.24. Fits to the data yield $\beta = 0.113(4)$ and $\beta = 0.114(4)$ using two different spectrometer collimations (Hagen and Paul, 1984). Similar results were obtained using K_2CoF_4 (Cowley *et al.*, 1984a). The experimental values for β are in fair

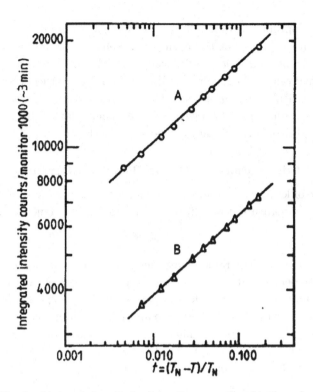

Fig. 4.24: The $d = 2$ integrated scattering intensity, proportional to the order parameter, versus t measured for Rb_2CoF_4 taken at two different reciprocal lattice points (Hagen and Paul, 1984). Reproduced with permission from IOP Publishing.

agreement with the exact value $\beta = 1/8$ for $d = 2$ and very different from the $d = 3$ Ising value $\beta = 0.33$ (see Table 2.1).

4.4.2 $d = 3$ *Pure Ising order parameter*

The $d = 3$ Ising order parameter is best determined experimentally in the highly anisotropic and well-characterized antiferromagnet FeF_2. For the antiferromagnetic systems such as FeF_2, with high crystalline quality and a concentration of the scattering at the $d = 3$ Bragg points, extinction effects preclude using neutron scattering as a tool for characterizing the critical behavior of the antiferromagnetic order parameter. The scattering is so strong that the beam gets depleted long before transversing the crystal. Fortunately, the order parameter can be measured using other techniques. The temperature dependence of the hyperfine effective field, H_{eff},

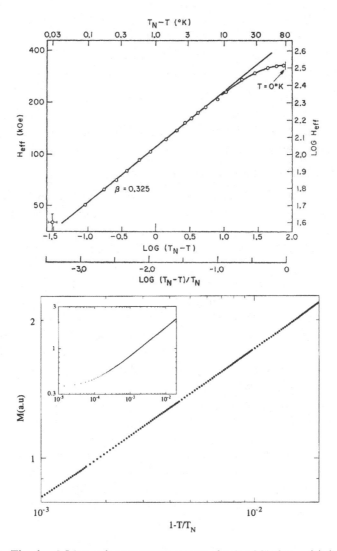

Fig. 4.25: The $d = 3$ Ising order parameter measured using Mössbauer (a) (Wertheim and Buchanan, 1967) and magnetometry (b) (Mattsson *et al.*, 1994) techniques. Reproduced with permission from American Physical Society (upper) and Elsevier (lower).

which is proportional to the magnetic order, has been determined from Fe[57] Mössbauer spectra (Wertheim and Buchanan, 1967) and is shown in Fig. 4.25 (upper panel), and the fit to the data gives the exponent $\beta = 0.325(5)$. The order-parameter exponent has also been determined to be $\beta = 0.325(2)$, as shown in Fig. 4.25 (lower panel) from measurements of the weak spontaneous magnetization below T_N, likely a piezomagnetic effect (Mattsson *et al.*, 1994). The experimental values for β using the various techniques are consistent and are in excellent agreement with theory and simulation results listed in Table 2.1.

4.5 Neutron Scattering Critical Line Shapes of Pure Ising Antiferromagnets

4.5.1 $d = 2$ *Pure Ising neutron scattering*

Because the magnetic interactions in K_2CoF_4 and Rb_2CoF_4 are orders of magnitude greater within the square lattice planes compared to those between the planes, they are almost ideal $d = 2$ systems. As was seen above, the magnetic specific heat of Rb_2CoF_4 showed the expected logarithmic behavior in $|t|$ for $10^{-3} < |t| < 10^{-1}$ and follows the exact solution for the $d = 2$ Ising model quite well over a wide temperature range. The excellent $d = 2$ Ising behavior provides an opportunity to study in detail the neutron scattering line shapes and measure the critical behavior of the $d = 2$ Ising universality class. This is particularly interesting to do for $d = 2$ because the critical behavior is far from mean-field behavior that holds for $d \geq 4$ in pure Ising systems, so deviations from mean-field line shapes should be clearly observed.

The scattering line shapes for the $d = 2$ Ising antiferromagnet K_2CoF_4 were measured (Cowley *et al.*, 1984a) and examples are shown in Figs. 4.26 and 4.27. The $d = 2$ scattering from the two-dimensional magnetic planes was measured at $(q,0,0.45)$ along the the scattering rod in reciprocal space. Although there is an exact solution for the $d = 2$ Ising critical behavior, the line shapes must be approximated. The scattering data were fit to the model scattering line shapes developed for $T > T_N$ (Fisher and Burford, 1967),

$$\chi(q) = A^+ \kappa^{-2+\eta} \frac{(1 + \phi_c^2 q^2 / \kappa^2)^{\eta/2}}{1 + \psi q^2 / \kappa^2}, \qquad (4.8)$$

and $T < T_N$ (Tarko and Fisher, 1975),

$$\chi(q) = A^- \kappa^{-2+\eta} \frac{(1 + \phi^2 q^2 / \kappa^2)^{\eta/2}}{(1 - \lambda + \lambda(1 + q^2/\kappa^2)^{1/2})^2}. \qquad (4.9)$$

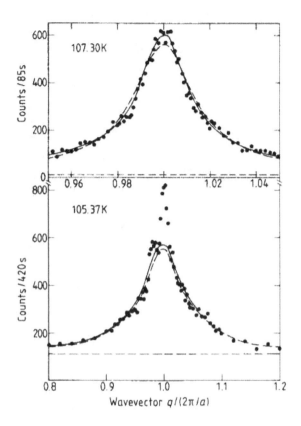

Fig. 4.26: Scattering line shapes for $T < T_N$ for K_2CoF_4 (Cowley *et al.*, 1984a).

The various parameters in these equations are listed in Table 4.3 and q is in units of $2\pi/a$ where a is the lattice parameter. These equations are improvements over the modified mean-field Ornstein–Zernike Lorentzian line shapes often used in analyzing neutron scattering data. For $d = 2$, reliable characterizations of the universal Ising critical parameters requires the use of the non-Lorentzian line shape for $T < T_N$, while a simple modification of the Ornstein–Zernike line shape (Tracy and McCoy, 1975) works fairly well for $T > T_N$ and $t > 10^{-3}$. Not only do the approximate line shapes above allow the extraction of universal critical parameters, they also serve as a warning not to put too much trust in critical parameters obtained using simple modified Lorentzian line shapes, as is a common practice when more correct line shapes are not available.

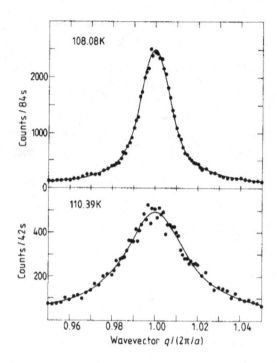

Fig. 4.27:   Scattering line shapes for $T > T_N$ for K_2CoF_4 (Cowley *et al.*, 1984a).

Table 4.3:   Parameters
for the line shapes of
the $d = 2$ Ising model
(Fisher   and   Burford,
1967; Tarko and Fisher,
1975).

η	0.25
ϕ_c	0.0294
ψ	$1 + \frac{1}{2}\eta\phi^2 = 1.001$
ϕ	0.2
λ	0.45

The values of various critical parameters obtained from the fits to the K_2CoF_4 (Cowley *et al.*, 1984a) data are listed in Table 4.4. The behavior of χ, for example, is shown in Fig. 4.28 (Cowley *et al.*, 1984a). The results are in good agreement with the values from theory. Also listed are the results and prediction for the two-scale universality parameter R_s.

Table 4.4: The measured $d = 2$ critical exponents from neutron scattering on K_2CoF_4 (Cowley et al., 1984a).

	$10^{-3} < |t| < 10^{-1}$
ν for $T > T_N$	1.02(5)
ν for $T < T_N$	1.12(13)
η	0.40(2)
κ^+/κ^-	1.85(22)
γ for $T > T_N$	1.82(7)
γ for $T < T_N$	1.92(20)
χ^+/χ^-	32.6(3.7)
β	0.155(2)
R_s	0.0565(75)

Note: The results compare fairly well overall with the values in Tables 2.1 and 2.2.

Two-scale universality (see Chapter 2) was tested in Rb_2CoF_4 which, like K_2CoF_4, is extremely close, but not perfectly so, to an ideal $d = 2$ Ising model. The data for $T > T_N$ were fit to a modified Lorentzian shown to be appropriate (Tracy and McCoy, 1975), similarly to the analysis in K_2CoF_4 above. The values of κ^+ significantly above T_N are shown in Fig. 4.29. The two sets of data are for two different spectrometer collimations; the set with tighter collimation (B) is offset vertically for clarity. The solid lines are fits to the usual asymptotic behavior $\kappa^+ = 1/\xi^+ = C^+ t^\nu$ and the dashed curves are fits to the behavior expected from the Onsager solution for the $d = 3$ Ising model on a square lattice, which should be accurate over a large range in temperature.

Fits to the neutron scattering data yield (Hagen and Paul, 1984) the order-parameter exponents $\beta = 0.114(2)$ and $0.113(2)$ for the two sets of spectrometer collimations used in the experiments. The resolution functions, which depend on the collimations, were determined as carefully as possible, though it was not as straightforward as measuring the Bragg scattering line widths, the method used for $d = 3$ described below. Because the order parameter, susceptibility, and correlation length critical behavior were measured from scattering data on a single crystal, and extinction affecting the order parameter is not an issue, it is possible to test two-scale-factor universality. The value obtained (Hagen and Paul, 1984) for Rb_2CoF_4, $R_s = 0.043(2)$, is somewhat lower than the expected (Tarko and Fisher, 1975) value 0.051, However, the experiments described above

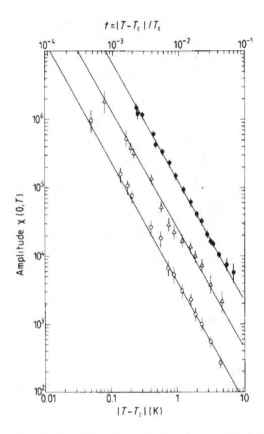

Fig. 4.28: Values of χ calculated from fits to the line shapes of K_2CoF_4 (Cowley *et al.*, 1984a). The upper set of results were obtained for $T > T_N$ using a Lorentzian line shape. The lower curve used the line shape in Eq. (4.9) for $T < T_N$, which gave a reasonable ratio for χ^+/χ^-. The results in the center were obtained using a Lorentzian line shape for $T < T_N$, which gave a reasonable exponent, but the ratio χ^+/χ^- is far from the theoretical expectation. Using the simple Lorentzian line shape can yield fairly good exponents, but not amplitude ratios.

(Cowley *et al.*, 1984a) yielded a value $R_s = 0.0565(75)$, which is higher than the predicted value by roughly the same amount. The discrepancies are likely due to the difficulty in correcting for the significant instrumental resolution for the scattering that takes place along the rods in reciprocal space, and the two experiments can be taken together as good evidence that two-scale-factor universality holds as expected.

Overall, the neutron scattering for the $d = 2$ Ising system is consistent with theoretical expectations. However, the amplitude ratios are not

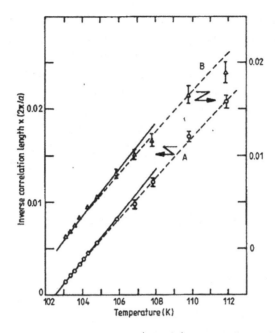

Fig. 4.29: The inverse correlation length, $\kappa^+ = 1/\xi^+$ versus T for $T > T_C$ for the $d = 2$ Ising antiferromagnet Rb_2CoF_4 (Hagen and Paul, 1984). The dataset B has been offset. The dashed curves are the expected behavior from the exact $d = 2$ Ising solution while the solid curves are the power law behaviors. Reproduced with permission from IOP Publishing.

accurately determined with Lorentzian line shape analyses and the theoretical line shape expressions are essential for that purpose. The need for good line shape models will be a theme for all $d = 2$ and $d = 3$ critical behavior studies using neutron scattering techniques.

4.5.2 $d = 3$ *Pure Ising neutron scattering*

In the previous section, the non-Lorentzian character of the $d = 2$ scattering line shapes was shown to be particularly significant for $T < T_N$. The data below the transition could not be adequately fit without the more accurate theoretical non-Lorentzian line shapes developed. The non-Lorentzian character of the scattering line shapes for $d = 3$ are more subtle than the case of $d = 2$ because the critical behavior is closer to mean-field behavior that holds for $d \geq 4$. For $d = 3$, the modified Lorentzian approximation

$$\chi(q) = \frac{A\kappa^\eta}{q^2 + \kappa^2}, \tag{4.10}$$

typically works fairly well, but only for $|t| > 10^{-3}$. To probe the asymptotic critical behavior closer to T_N, the theoretical line shapes that were developed for the $d = 3$ Ising model were used. The line shapes were tested using FeF$_2$ and were found to work well; power law behavior described the behavior of κ and χ for $|t| > 10^{-4}$, nearly matching the range of $|t|$ used in the specific heat characterizations. Furthermore, the line shapes that were developed for $d = 3$ also do a better job of yielding critical amplitudes that are consistent for data across the critical region. The line shapes are (Fisher and Burford, 1967; Martín-Mayor *et al.*, 2002; Tarko and Fisher, 1975)

$$S(q) = A^+ \kappa^{-2+\eta} \frac{(1 + \phi^2 q^2 / \kappa^2)^{\eta/2}}{1 + \psi q^2 / \kappa^2}, \tag{4.11}$$

for $T > T_N$ and

$$S(q) = A^- \kappa^{-2+\eta} \frac{(1 + \phi'^2 q^2 / \kappa^2)^{\sigma+\eta/2}}{(1 + \psi' q^2 / \kappa^2)(1 + \phi''^2 q^2 / \kappa^2)^\sigma}, \tag{4.12}$$

for $T < T_N$. The parameters for $d = 3$ are given in Table 4.5.

To test the line shapes and extract universal critical parameters from fits to scattering data, FeF$_2$ was chosen because it is a strongly anisotropic antiferromagnet with magnetic and structural properties that are well characterized. In addition, the specific heat and order parameter critical behaviors have been measured and agree with theoretical values. The neutron scattering line shape data were taken as close as $|t| = 10^{-4}$, which should ensure that the measurements are within the asymptotic critical region and deviations from mean-field line shapes should be most apparent for small $|t|$. It should be noted that crossover to asymptotic critical behavior takes

Table 4.5: Parameters used for the line shapes in Eqs. (4.11) and (4.12) (Fisher and Burford, 1967; Tarko and Fisher, 1975).

η	$\frac{1}{18} = 0.056$
ϕ	0.15
ψ	$1 + \frac{1}{2}\eta\phi^2 = 1.001$
σ	$2\eta = 0.111$
ϕ'	0.3247
ϕ''	0.09355
ψ'	$1 + \frac{1}{2}\eta\phi'^2 + \sigma(\phi'^2 - \phi''^2) = 1.0137$

place as both $|t|$ and $|q|$ decrease. However, for large values of $|q|$, the scattering intensity becomes insignificant, so it is assumed that the main effect of crossover should be seen as $|t|$ decreases and $|q|$ is much less important.

As an aside, for these measurements, as well as those done on the $d = 2$ experiments on K_2CoF_4, the resistance bridges available at the time did not have the resolution needed to measure the temperature accurately enough with platinum thermometers to do the critical behavior measurements. However, two bridges were used, one to measure the resistance of the platinum thermometer and another to measure the resistance of a thermistor. The resistances of both thermometers were recorded for each line shape measurement. The thermistor provided the necessary temperature resolution while the platinum thermometer was used to calibrate the thermistor. The stability in the case of FeF_2 was enhanced by using pumped liquid nitrogen as the cryogenic coolant, so the environment around the sample was always near to the temperature of the sample. The combination of thermometers and cryogenic coolant allowed measurements at reduced temperatures as small as $|t| = 10^{-4}$.

Because the differences between the Lorentzian and non-Lorentzian line shapes are small, the significant corrections for instrumental resolution were made carefully. Unlike the $d = 2$ experiments, the instrumental resolution is well represented by an ellipsoid with axes along the vertical, longitudinal and transverse scattering directions in reciprocal space. The three measured resolution line shapes, taken at low temperature where magnetic fluctuations are negligible, are shown in Fig. 4.30. The scattering plane is horizontal and the vertical resolution is the widest. The resolution is most narrow in the transverse direction that will be used for the scans that are used to test the line shapes. The Bragg scattering data could be fit to Gaussian line shapes and the widths could then be used to integrate over the resolution ellipsoid to obtain resolution corrected data for the line shapes at $|q| > 0$. However, the data for $|q| > 0$ were instead corrected using the resolution data directly with the equation

$$I(q) = I_0 \frac{\sum_a \sum_b \sum_c S((q - q_0 - a)^2 + b^2 + c^2) T_{(a)} L_{(b)} V_{(c)}}{\sum_a \sum_b \sum_c T_{(a)} L_{(b)} V_{(c)}} + B \quad (4.13)$$

as a part of the line shape fitting.

Figure 4.31 shows two typical line shapes for $T < T_N$ and fits to the data. The fits of the scattering data at each temperature to the line shapes give the inverse correlation length κ and the susceptibility χ at those temperatures, as shown in Figs. 4.32 (upper panel) and 4.33, respectively.

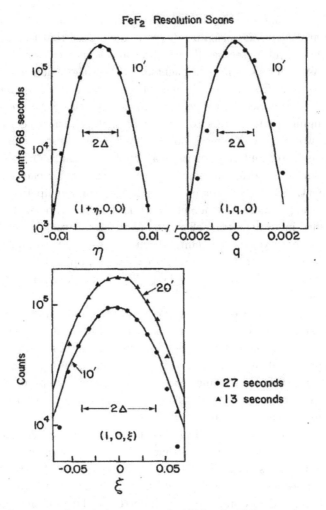

Fig. 4.30: Instrumental resolution scans at low T for the FeF_2 measurements and fits to Gaussian line shapes (Belanger and Yoshizawa, 1987).

The values of the critical exponents γ and ν obtained from χ and κ versus t are shown in Table 4.6. The experimental values are in excellent agreement with the predicted values.

If the Lorentzian line shapes are used in the analysis, the power law behavior of χ and κ is distorted for $|t| < 10^{-3}$. This is demonstrated in Fig. 4.32. The fitted values curve upward from the straight-line behavior at larger $|t|$. To demonstrate that it is a consequence of using the incorrect line

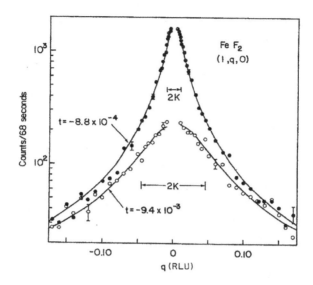

Fig. 4.31: FeF_2 scattering line shape data for $T < T_N$ and fits to Eq. (4.12) (Belanger and Yoshizawa, 1987). The Bragg scattering data have been excluded from the plot and the fits.

shapes, line shapes were generated using Eqs. (4.11) and (4.12). Random noise was added and the line shapes were fit using the Lorentzian line shapes. The solid curves are fits to those generated data and follow the real results obtained from the Lorentzian fits quite well.

Not all is well for $|t| > 10^{-3}$ using the Lorentzian line shapes. Although for $d = 3$, the values of κ and χ are largely insensitive to whether Eq. (4.10) or Eqs. (4.11) and (4.12) are used in the data fitting except for $|t| < 10^{-3}$, the amplitudes A^+/A^- are sensitive over the entire critical region. The comparison of the amplitudes obtained using the different line shapes is shown in Fig. 4.34. This serves as a cautionary note that when line shapes are not known accurately, fitted exponents might be fairly accurate while the fitted amplitudes might be quite inaccurate. This is a serious concern for the random-field Ising model studies, where there is little guidance from theory or computer simulations regarding line shapes, as discussed in Section 4.7.2.

Unlike the $d = 2$ experiments, the $d = 3$ Bragg scattering is concentrated at points in reciprocal space and is high in intensity. The extremely high crystalline quality of FeF_2 samples leads to extinction effects that preclude the use of the temperature dependence of the Bragg scattering at

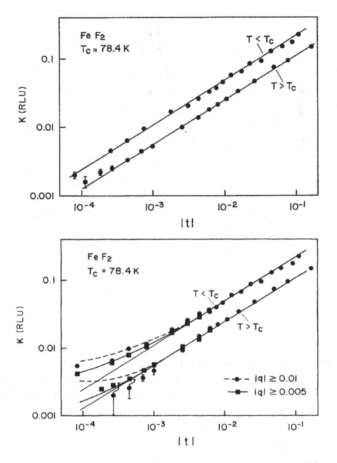

Fig. 4.32: The critical behavior of κ from analysis of the line shapes. The upper panel shows power law behavior obtained using Eqs. (4.11) and (4.12). The lower panel shows the nonpower law behavior, especially for $T < T_N$ obtained from fits using Eq. (4.10) (Belanger and Yoshizawa, 1987). The nonlinear curves are simulated behaviors as described in the text.

$q = 0$ to characterize the order-parameter critical behavior. The techniques described in Section 4.4 are better suited to that purpose.

Just as we saw for $d = 2$, the exponents are reasonably well characterized with Lorentzian line shapes using data not too close to T_N, but the amplitudes need the more accurate line shape models from theory. The theoretical line shapes prove to be helpful in accurately characterizing the critical behavior close to T_N. It is not always practical to analyze data

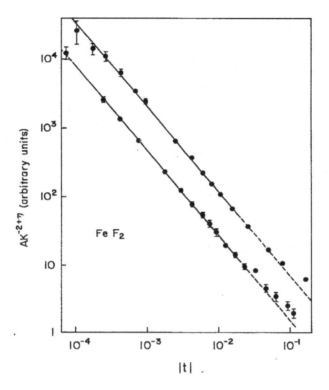

Fig. 4.33: The values of χ obtained using Eqs. (4.11) and (4.12) (Belanger and Yoshizawa, 1987).

Table 4.6: Critical exponents and amplitude ratios obtained from fits to FeF_2 data (Belanger and Yoshizawa, 1987).

	$10^{-4} < |t| < 10^{-1}$
ν	0.65(1)
$\kappa^+/\kappa^- = \xi^-/\xi^+$	0.54(1)
γ	1.26(1)
χ^+/χ^-	4.6(1)

using the non-Lorentzian line shapes, and they are not yet available for many models. It is important to be aware of the limitations on the accuracy of the critical behavior characterizations that are done without the more accurate line shapes.

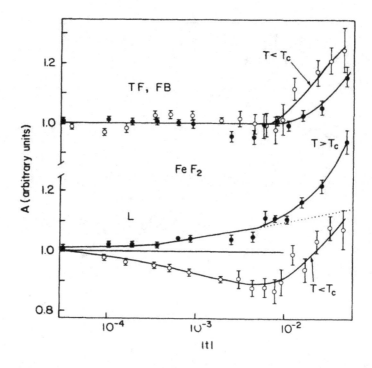

Fig. 4.34: The behavior of the line shape amplitudes obtained from the fits. Note that using Eqs. (4.12) and (4.11) yields equal amplitudes for $T > T_N$ and $T < T_N$ over a wide critical region, whereas using Eq. (4.10) shows the amplitudes varying significantly over the same range (Belanger and Yoshizawa, 1987).

4.6   The Random-Exchange Ising Model

The universal critical behavior of a pure magnetic system can be altered upon introduction of quenched random dilution, but only when the specific heat exponent of the pure system is positive. The scaling argument that was used to predict the change is known as the Harris criterion (Harris, 1974). For cases where α is already negative in the pure case, the dilution is not expected to alter the critical behavior. For the $d = 3$ Ising case, where $\alpha > 0$ for the pure case, the random-exchange exponent α should become negative as the critical behavior crosses over from pure to random-exchange critical behavior as $|t|$ becomes smaller. Along with the change in sign of α, the amplitude ratio A^+/A^- should change from the pure case, where $A^+/A^- < 1$, to $A^+/A^- > 1$. This has been shown (Birgeneau *et al.*, 1983a; Slanič and Belanger, 1998) to be the case for $Fe_x Zn_{1-x} F_2$, where the exponents, and corresponding amplitude ratios, resemble those

Table 4.7: The $d = 3$ pure and random-exchange specific heat critical behaviors characterized using $d(\Delta n)/dT$ versus T for the pure FeF_2 and magnetically dilute $d = 3$ $Fe_xZn_{1-x}F_2$ Ising antiferromagnets.

α		
Pure	0.111(7),0.114(4)	Belanger *et al.* (1983b)
RE ($x = 0.93$)	−0.10(2)	Slanič and Belanger (1998)
RE ($x = 0.6$)	−0.09(3)	Birgeneau *et al.* (1983a)
A^+/A^-		
Pure	0.543(20),0.528(10)	Belanger *et al.* (1983b)
RE ($x = 0.6$)	1.6(3)	Birgeneau *et al.* (1983a)

Notes: Because $\alpha > 1$ for the pure system, it must change sign for the dilute system and that is what is observed. The random-exchange behaviors were all obtained using the birefringence technique.

of the $d = 3$ pure Heisenberg values, with α negative and $A^+/A^- > 1$. As demonstrated in Table 4.7, the new critical behavior of the specific heat does not appear to depend on x. When α changes from being positive to being negative, the new critical behavior should exhibit changes in the other universal exponents and amplitude ratios as well. The specific heat is discussed in detail in Section 4.6.3.

In this discussion, we will cover antiferromagnetic systems that have randomness in their exchange interactions, specifically when each site is randomly magnetic or nonmagnetic. As long as the concentration of interactions is greater than the percolation threshold (Stauffer and Aharony, 1994), which is $x_p = 0.246$ for the $d = 3$ fluorides on a body-centered tetragonal lattice, or $x_p = 0.593$ for the square lattice fluorides, a transition can occur. The critical behavior will belong to a universality class different from the pure case for the $d = 3$ Ising systems. The predicted exponent values are listed in Table 4.8, and they can be compared to the pure $d = 3$ Ising exponents in Table 2.1. The $d = 2$ Ising case is marginal because $\alpha = 0$. As we will see below, there is a striking difference between the pure and dilute systems beyond changes in the static critical behaviors. The dilute systems can take much longer to equilibrate and this will become apparent in the dynamical critical behavior as well as at low temperatures.

4.6.1 $d = 2$ *Random-exchange Ising model*

Whereas the specific heat exponent α for the $d = 3$ Ising model changes from positive to negative upon dilution, in accordance with the Harris criterion, the logarithmic specific heat peak of the $d = 2$ Ising model is not expected to exhibit such a dramatic change because, for the pure case,

Table 4.8: Theoretical estimates for the critical exponents of the $d = 3$ random-exchange Ising model.

γ	1.31–1.34
ν	0.67–0.70
η	0.03–0.05
α	(-0.09)–(-0.01)
β	0.34–0.35

Note: The values represent a reasonable range of the extensive list compiled by Pelissetto and Vicari (2002).

$\alpha = 0$. Corrections to the critical behavior are expected (Dotsenko and Dotsenko, 1982) but for experimentally accessible ranges of $|t|$ the specific heat of the dilute system will be $C \propto \ln(|t|)$ for small $|t|$, which is also the behavior of the pure system. Indeed, the observed behavior (Ferreira *et al.*, 1983) in the magnetically diluted system is similar in appearance to the pure case for small values of $|t|$, as shown in Figs. 4.35 and 4.36. Using

$$C = -A^{\pm} \ln(|t|) + B + Et, \tag{4.14}$$

where B and Et are background terms, where A^+ and A^- are the amplitudes above and below T_N, respectively, the amplitude ratio $A^+/A^- = 0.95(10)$ was obtained from fits to the data.

A neutron scattering study (Ikeda *et al.*, 1979) in the $d = 2$ pure and random-exchange antiferromagnets $Rb_2Co_xMg_{1-x}F_4$, with $0.89 \leq x \leq 1$ yielded the exponents $\nu = 0.99(4)$ and $1.05(0)$ for $x = 1$ and 0.89, respectively, and $\gamma = 1.67(9)$ and $1.67(8)$ for $x = 1$ and 0.89, respectively, using the line shape

$$S(q) = \frac{A}{(\kappa^2 + q^2)^{1-\eta/2}}, \tag{4.15}$$

for $T > T_N$. Because there is no guidance regarding the line shapes expected for random-exchange systems analogous to the pure case, simple modifications to mean-field expressions were used. The results for κ and χ versus t above the transition temperature are shown in Fig. 4.37. As expected, the critical behavior does not appear to change with dilution.

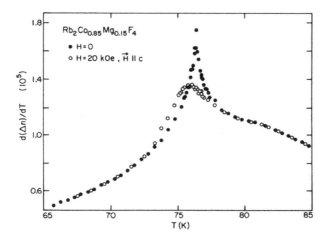

Fig. 4.35: The specific heat of the $d = 2$ magnetically dilute $Rb_2Co_{0.85}Mg_{0.15}F_4$ Ising system measured using the $d(\Delta n)/dT$ technique (Ferreira *et al.*, 1983). Also shown is the magnetic specific heat measured in a small applied field, which corresponds to the destroyed $d = 2$ random-field Ising model transition that will be discussed in Section 4.7.1. Reproduced with permission from American Physical Society.

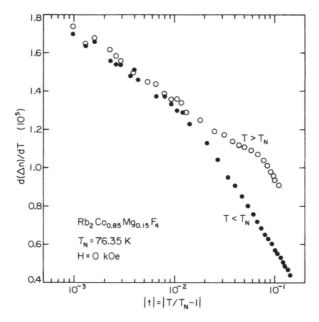

Fig. 4.36: Semi-log plot of the $d = 2$ magnetically dilute $Rb_2Co_{0.85}Mg_{0.15}F_4$ Ising systems measured using the $d(\Delta n)/dT$ technique (Ferreira *et al.*, 1983). The data show an apparent symmetric logarithmic behavior similar to the pure $d = 2$ Ising system Rb_2CoF_4. Reproduced with permission from American Physical Society.

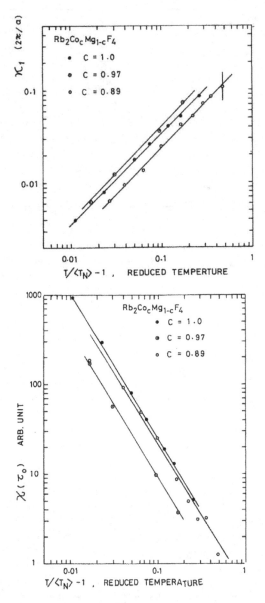

Fig. 4.37: The behavior of κ and χ determined from fits to the data from measurements on the $d = 2$ dilute Ising system $Rb_2Co_{0.85}Mg_{0.15}F_4$ (Ikeda *et al.*, 1979). Reproduced with permission from the Physical Society of Japan.

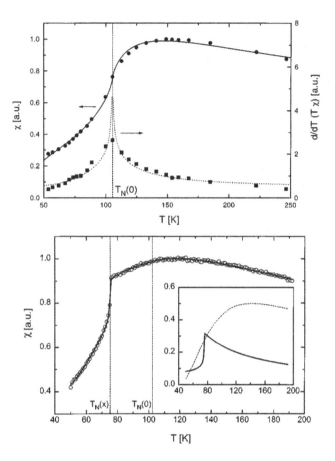

Fig. 4.38: The behavior of κ and χ of pure Rb_2CoF_4 (upper) and $Rb_2Co_{0.85}Mg_{0.15}F_4$ (lower) (Binek *et al.*, 1998). The influence of Griffiths singularities in the lower panel, as explained in the text, is between $T_N(0)$ and $T_N(x)$, as marked by the vertical lines.

Although the critical behavior is unchanged for $d = 2$ upon dilution, the behavior of the uniform magnetization is altered by the dilution between the transition temperature of the dilute system and the transition temperature of pure system (Binek *et al.*, 1998). The upper panel in Fig. 4.38 shows the susceptibility χ versus T measured (Breed *et al.*, 1969) in Rb_2CoF_4 as well as $d(T\chi)/dT$. The lower panel of Fig. 4.38 shows χ versus T for $Rb_2Co_{0.85}Mg_{0.15}F_4$. The inset is a decomposition of the behavior

$$\chi(T) = \frac{1}{T}(A_1 + A_2|t|^{1-\alpha} - A_3|t|^{2\beta+\gamma-\phi}), \tag{4.16}$$

where $t = (T - T_N)/T_N$, $\phi \approx 1.85$ is the measured $d = 2$ crossover exponent, and α, β, and γ are the random-exchange exponents, which for $d = 2$ are equal to the $d = 2$ pure exponents. The $A_3|t|^{2\beta + \gamma - \phi}$ term is absent in the pure Rb_2CoF_4 case. Its form was predicted for the critical region (Aharony, 1986; Fishman and Aharony, 1979). That it appears well outside the critical region is attributed to Griffiths singularities (Griffiths, 1969) that take place between the dilute and pure transition temperatures.

4.6.2 $d = 3$ *Random-exchange Ising order parameter*

The random-exchange order-parameter exponent $\beta = 0.350(9)$ has been measured using Mössbauer spectroscopy (Rosov *et al.*, 1988) for the $d = 3$ random-exchange Ising system $Fe_{0.9}Zn_{0.1}F_2$ as shown in Fig. 4.39. Other measurements including $\beta = 0.35(1)$ using anomalous dilation (Ramos *et al.*, 1988b) and $\beta = 0.348(10)$ using using NMR (Sartorelli, 1992). As will be discussed below, X-ray scattering can be used to measure the order parameter without the effects of extinction. The random-exchange order parameter $\beta = 0.35(2)$ was measured (Ye *et al.*, 2006, 2002) in $Fe_{0.85}Zn_{0.15}F_2$. All of these measured values are consistent with each other and with the predicted value range in Table 4.8. The value of β for the $d = 3$ random-exchange Ising is slightly larger than the pure $d = 3$ Ising model value listed in Table 2.1. The change is consistent with the expectation

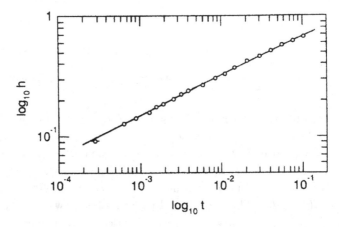

Fig. 4.39: The $d = 3$ Ising random-exchange order-parameter measured in $Fe_{0.9}Zn_{0.1}F_2$ using Mössbauer spectroscopy (Rosov *et al.*, 1988). Reproduced with permission from American Physical Society.

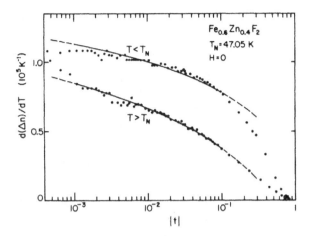

Fig. 4.40: A semi-log plot of the magnetic specific heat of $Fe_{0.6}Zn_{0.4}F_2$ obtained using the birefringence technique (Birgeneau *et al.*, 1983a).

(Harris, 1974) that the critical behavior upon magnetic dilution changes when the pure specific heat exponent, α, is greater than zero.

4.6.3 $d = 3$ *Random-exchange Ising specific heat critical behavior*

The magnetic specific heat of the $d = 3$ random-exchange Ising system $Fe_{0.6}Zn_{0.4}F_2$ in Fig. 4.40, measured using the birefringence technique, shows the negative curvature throughout the critical region (Birgeneau *et al.*, 1983a). A fit to the data gives the negative random-exchange value $\alpha = -0.09(3)$, contrasting the positive value obtained using FeF_2. That result is obtained for a magnetic concentration $x < x_v$, where x_v is the percolation threshold concentration for vacancies. A similar result, $\alpha = -0.10(2)$ (Slanič and Belanger, 1998), was obtained from a fit to data with $x = 0.93 > x_v$. The random-exchange specific heat results for the exponents and amplitudes are listed in Table 4.7 along with those from the pure system.

Using the scaling relation for the crossover from pure to random-exchange critical behavior upon dilution,

$$F(t, 1 - x) = |t|^{2-\alpha} f((1 - x)/t^{\phi}), \qquad (4.17)$$

it was shown (Ferreira *et al.*, 1991a) that the specific heat critical amplitudes should scale as $(1 - x)^{(\alpha_{re} - \alpha)/\phi}$, where α and α_{re} are the pure and

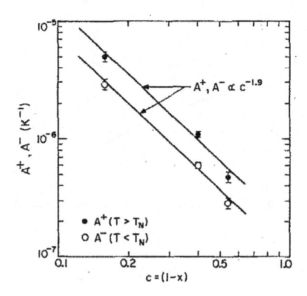

Fig. 4.41: A log–log plot of the concentration dependence of the specific heat critical behavior amplitudes in $Fe_x Zn_{1-x} F_2$ determined from the data shown in Fig. 3.5 (Ferreira *et al.*, 1991a). The solid lines show the expected $1 - x$ dependence. Reproduced with permission from AIP Publishing.

random-exchange specific heat exponents and ϕ is the appropriate crossover exponent; in this case, $\phi = \alpha$. Analyzing the data from Fig. 3.5, Fig. 4.41 shows that expected concentration dependence, $(1 - x)^{1.9}$, is exhibited for both A^+ and A^- as a function of $(1 - x)$. This once again supports the proportionality between the specific heat and $d(\Delta n)/dT$.

4.6.4 $d = 3$ Random-exchange neutron scattering

The characterization of the $d = 3$ random-exchange critical behavior of χ and κ using neutron scattering line shapes in the dilute anisotropic antiferromagnet $Fe_x Zn_{1-x} F_2$ presents challenges not encountered in experiments using the pure system FeF_2 system. First, because the neutron experiments require the use of relatively large crystals, the concentration gradients must be minimal in order to closely approach the transition temperature. Fortunately, a crystals with small concentration gradients were produced with $x = 0.46$ (King *et al.*, 1988), $x = 0.85$ and $x = 0.93$. Second, there is little guidance from theory regarding the expected line shapes for random-exchange systems, unlike those developed for pure systems (Fisher and Burford, 1967; Tarko and Fisher, 1975). In the pure systems, Lorentzian line shapes modified slightly to ensure proper scaling limits were adequate for $|t| > 10^{-3}$. Since it is difficult in the magnetically dilute system to approach the transition closer than this, that might be adequate here as well. However, the most reliable universal parameter estimates, especially the amplitude ratios, were obtained in the pure system with more accurate line shapes (see Section 4.5.2).

The neutron scattering data obtained with the crystal with $x = 0.46$ were analyzed using modified Lorentzian line shapes and the results for the critical parameters are shown in Table 4.9. These can be combined with the specific heat critical behavior parameters obtained with the same crystal to test the hyperscaling relation in Eq. (2.11). Rearranging, we have $d\nu + \alpha = 2$. For $x = 0.46$, this gives $2.07(3) - 0.09(3) = 1.98(4)$ and for the non-Lorentzian fit of the $x = 0.93$ data, this gives $2.10(6) - 0.10(2) = 2.00(7)$, so hyperscaling is satisfied for the $d = 3$ random-exchange Ising exponents.

Although scaling relations are satisfied using the parameters obtained from fits to the Lorentzian line shape, it is unclear from that analysis what changes might occur if data closer to T_N are analyzed using more accurate line shapes. Furthermore, at high magnetic concentrations, data closer to T_N are essential to avoid issues with pure-to-random exchange crossover. To study $x = 0.93$, Lorentzian line shapes also proved adequate, even close to T_N and the results are shown in Table 4.9. The results are similar to the $x = 0.46$ sample, but it is desirable to do the analysis with better line shapes to show whether the Lorentzian line shapes are yielding reliable results. Absent guidance from theory, the same line shapes that were used with FeF$_2$ were applied to the Fe$_{0.93}$Zn$_{0.07}$F$_2$ system, but with parameters σ, ϕ, ϕ', and ϕ'' allowed to vary in the fits. Unfortunately, there are too many parameters to meaningfully fit the data for any given set of data at

Table 4.9: $d = 3$ Random-exchange critical parameters (Belanger *et al.*, 1986; Slanič, 1998; Slanič *et al.*, 2001).

	$x = 0.46$	$x = 0.93$ Lorentzian	$x = 0.93$ non-Lorentzian
$|t|_{min}$	1.5×10^{-3}	1.1×10^{-4}	1.1×10^{-4}
$|t|_{max}$	10^{-1}	10^{-2}	10^{-2}
ν	0.69(1)	0.70(3)	0.70(2)
κ^+/κ^-	0.69(1)	0.47(4)	0.50(3)
γ	1.31(3)	1.34(3)	1.34(2)
χ^+/χ^-	2.8(2)	4.86(80)	4.47(52)
η		0.079(10)	0.079(12)

Table 4.10: $d = 3$ Random-exchange line shape parameters (Belanger *et al.*, 1986; Slanič *et al.*, 2001).

σ	0.16(20)
ϕ	0.18(2)
ϕ'	0.18(2)
ϕ''	0.08(10)

one temperature. Instead, all of the data for $T < T_N$ were fit together and all of data for $T > T_N$ were fit together. That provided enough input to achieve good fits, as shown by the universal parameters in Table 4.9. The line shape parameters obtained are given in Table 4.10.

The critical behavior exponents obtained for $x = 0.93$ with the non-Lorentzian line shapes are nearly the same as those from fits with the Lorentzian ones. They are also fairly consistent with results of earlier studies with $x = 0.46$ (Belanger *et al.*, 1986) except for the amplitude ratios; this could be a result of the analysis being done further from T_N for $x = 0.46$, where the value of χ^- extracted from the scattering line shapes below T_N could be affected (Pelcovits and Aharony, 1985). Furthermore, the parameters for the non-Lorentzian line shapes and the critical exponents are not that different from those specified by theory for FeF_2. It is clear that the Lorentzian line shapes work fairly well for characterizing the dilute $d = 3$ Ising model systems with $H = 0$ fairly close to T_N.

4.6.5 $d = 3$ *Random-exchange dynamics*

Random-exchange experiments exhibit slow, but seemingly conventional dynamics in the $Fe_{0.46}Zn_{0.54}F_2$ (Barrett, 1986; Belanger *et al.*, 1988b) and $Fe_{0.9}Zn_{0.1}F_2$ (Hohenemser *et al.*, 1989). Spin-echo neutron scattering techniques were used to study the critical dynamics of the random-exchange system $Fe_{0.46}Zn_{0.54}F_2$ with $H = 0$ and the value $z = 1.7(2)$ was obtained (Belanger *et al.*, 1988b). The value $z = 1.5$ was obtained (Barrett, 1986) using Mössbauer techniques at the same concentration $x = 0.46$. The exponent determined by Mössbauer techniques at the magnetic concentration $x = 0.9$ is much higher, with $z = 2.18(10)$. Figure 4.42 shows, for $x = 0.46$, the much slower dynamics in the random-exchange system $Fe_{0.46}Zn_{0.54}F_2$. A theoretical result for the random-exchange value (Hasenbusch *et al.*, 2007), $z = 2.35(2)$, which is higher than the pure $d = 3$ Ising exponent

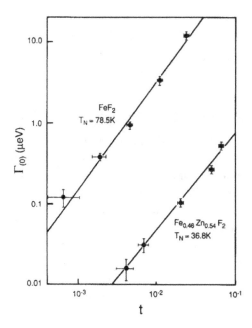

Fig. 4.42: A log–log plot of the Γ versus t for FeF_2 and $Fe_{0.46}Zn_{0.54}F_2$ obtained using neutron spin-echo techniques (Belanger *et al.*, 1988b). Γ is proportional to $\xi^{-z} = t^{\nu z}$.

(Wansleben and Landau, 1987) $z = 2.03(4)$, suggests that all of the experimentally determined values are low and that the discrepancy grows with magnetic dilution. The discrepancy at $x = 0.9$ has been attributed (Ballesteros *et al.*, 1998) to a strong crossover from pure to random-exchange dynamic critical behavior. For $x = 0.46$, which is far below the threshold concentration for vacancy percolation, $x_v = 0.76$, might also be affected by crossover. The percolating vacancy lattice, which weakens the magnetic lattice, could affect the critical dynamics as well.

4.6.6 $d = 3$ *Random-exchange susceptibility*

The scaling relations for the linear and nonlinear susceptibilities in the limit $H \to 0$

$$\chi_L \propto |t|^{2-\alpha-\phi}, \tag{4.18}$$

and

$$\chi_{NL} \propto |t|^{2-\alpha-2\phi} \tag{4.19}$$

to show that, near T_N, FeCl$_2$ is consistent with $d = 3$ Ising values of α and ϕ (Kushauer and Kleemann, 1995), whereas magnetically dilute Fe$_{0.7}$Mg$_{0.3}$Cl$_2$ yields exponents consistent with the $d = 3$ random-exchange Ising model (Leitão and Kleemann, 1988).

4.7 The Random-Field Ising Model

The random-exchange Ising system in the previous section is a study of the effects of randomness in the exchange interaction between spins. The randomness is important for $d = 3$ because the pure Ising system has a positive specific heat critical exponent, α, whereas the random-exchange Ising system must have a negative value for α. However, there is no intrinsic frustration. The two sublattices have no conflict in acquiring long-range antiferromagnetic order other than slow dynamics relative to the pure system.

In the random-field Ising system, on the other hand, the randomness couples directly with the order parameter. The competition between the random-field and the long-range order is severe. For $d = 2$, the frustration of the order is so strong that long-range order cannot occur. The transition for no random fields is "destroyed" when the random field is introduced into the system. As discussed below, the transition for $d = 3$ persists with random fields that are not too large, but the nature of the transition is greatly altered. The universality class of the $d = 3$ random-field Ising model is distinct from the random-exchange and pure university classes. The interactions between theorists and experimentalists were intense in an effort to characterizes the new critical behavior of this difficult to study transition. The extremely slow activated dynamics of the $d = 3$ random-field Ising model has made that characterization a particularly challenging.

Although there have been reviews of the random-field Ising model, many such reviews do not focus on experimental work. However, one review (Jaccarino and King, 1990) accurately assessed the experimental situation three decades ago. The review here will cover some of the same experiments as well as the experimental work that has been done since then. One lesson emphasized in the earlier review is that the study of the random-field Ising model can be treacherous and a good comprehensive understanding of it must come from many studies using many different techniques. A good overview will necessarily take into account a large body of experimental work, as well as the theories most pertinent to that work.

In the random-field Ising model (Imry and Ma, 1975), a field fluctuates in direction and, perhaps, magnitude along the spin-ordering axis. The average field is zero on long length scales and the interactions between spins are short-ranged. A typical Hamiltonian is

$$H = \sum_{ij} J S_i S_j - \sum_i h_{rf} S_i, \qquad (4.20)$$

where h_{rf} is a random field applied to a uniform magnet with the average $\langle h_{rf} \rangle$ equal to zero. For a dilute Ising antiferromagnet, a simple Hamiltonian is

$$H = \sum_{ij} \epsilon_i \epsilon_j J S_i S_j - \sum_i \epsilon_i H S_i, \qquad (4.21)$$

where $\epsilon_i = 1$ when site i is occupied by a magnetic ion and $\epsilon_i = 0$ when it is occupied by a nonmagnetic ion. The effective random field also averages to zero. A cartoon depiction of a $d = 2$ Ising lattice with random fields applied is shown in Fig. 4.43 (left side). The tendency of spins to align with the local fluctuating field competes with the long-range ordering of the spins. A random field that fluctuates on the length scale of the lattice spacing is difficult to create in the laboratory. When it was shown (Cardy, 1984; Fishman and Aharony, 1979), however, that a random field could be introduced into a dilute anisotropic antiferromagnet by application of a uniform field, the possibility of experiments that probe the random-field Ising model became a reality. Figure 4.43 (right side) illustrates $d = 2$ dilute antiferromagnet with a uniform field applied. In some local regions, the

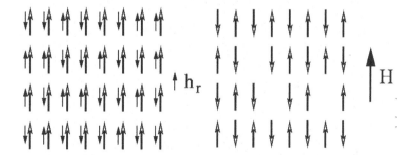

Fig. 4.43: A uniform $d = 2$ Ising ferromagnet with an applied random field h_r (left) and a $d = 2$ dilute Ising antiferromagnet with an applied uniform field (right). The random field h_r averages to zero over the lattice. The effective random field in the dilute antiferromagnet also averages to zero. Both the random field in the uniform magnet and the effective field generated by the uniform field in the dilute antiferromagnet compete with long-range ordering. For $d = 2$, the transition is destroyed by the random field, whereas in $d = 3$ the transition is governed by a new random-field Ising universality class.

random fluctuation in the number of spins on one sublattice favor alignment of that sublattice with the field. In other local regions, the other sublattice will have a majority of the spins and will be favored to align with the field. The locally varying preference in sublattice alignment with the field directly competes with long-range order in which each sublattice aligns globally in one direction opposite of the other. The effective random-field averages to zero and the occupation of lattice sites is random on short length scales. It has been shown that the phase transition universality class is the same for the two cases, a uniform magnet with a random field and a randomly dilute magnet with a uniform field (Cardy, 1984; Picco and Sourlas, 2015). A particularly important property of the experimental systems is that the random-field strength can easily be varied or even shut off simply by varying the uniform applied field. This allows the field to be a tuning parameter.

For $d = 2$, a FC domain state below the destroyed transition has been simulated (Nowak and Usadel, 1992a,b) as shown in Fig. 4.44. The black

Fig. 4.44:   A frozen FC domain state on a 399×398 size lattice in a dilute antiferromagnet (Nowak and Usadel, 1992a). The black and gray sites are magnetic and the white sites are vacancies. Reproduced with permission from Elsevier.

and grey sites are magnetic with different sublattices pointing along the applied field. The white sites are vacancies. Note that the black state has relatively small domains whereas the gray state is much more connected. Domain walls tend to pass through vacancies to reduce the energy of the walls.

For low enough dimensions, random fields destabilize long-range order. For example, for $d = 2$, Ising magnets order without random-fields, as shown earlier in this chapter. A basic argument (Imry and Ma, 1975) for the lower critical dimension d_l, below which long-range order cannot occur, is based on the balance of the energy advantage of locally aligning moments along the random field and the resulting energy cost of forming domain walls. The free energy for compact domains is given by (Imry and Ma, 1975) $W = gR^{d-1} - hm_0R^{d/2}$, where g is the domain wall surface tension, h^2 is the variance of the random field strength, and m_0 is the $H = 0$ moment at each magnetic site. For $d > 2$, it is energetically disadvantageous to form domains for small fields, implying $d_l = 2$ using this relationship. Ground state calculations show that domains are not compact (Esser *et al.*, 1997), and perturbation theory indicated that $d_l = 3$ (Aharony *et al.*, 1976; Young, 1977), but the lower critical dimension was in fact demonstrated to be $d_l = 2$ in equilibrium (Aizenman and Wehr, 1989; Bricmont and Kupiainen, 1987; Imbrie, 1984).

The behavior for $d = 3$ observed in experiments on random-field systems realized by applying uniform fields to dilute antiferromagnets is complex. Near the $d = 3$ random-field transition, hysteresis plays an important role in all high resolution experiments.

Among the most comprehensively studied families are the highly anisotropic body-centered tetragonal $Fe_xZn_{1-x}F_2$ antiferromagnet, the weakly anisotropic, isomorphic $Mn_xZn_{1-x}F_2$ antiferromagnet, and the layered antiferromagnet $Fe_xMg_{1-x}Cl_2$ for $d = 3$, and the highly anisotropic antiferromagnets $Rb_2Co_xMg_{1-x}F_4$ and $K_2Co_xMg_{1-x}F_4$ for $d = 2$. The exchange interactions in these systems are short-ranged and well characterized. They all can be grown with fairly uniform dilution (although this can be challenging) and excellent crystalline structural quality. The zero field random-exchange behaviors are well characterized in these systems, so the crossover to random-field behavior with the application of the field can be confidently characterized.

Of the phase transitions covered in this book, the characterization of the critical behavior of the $d = 3$ random-field Ising model has been, by far, the most difficult. Although progress has been made, many aspects of the theory, computer simulations and experiments remain elusive. The destruction of the $d = 2$ transition was well established early in experiments and theory. Early work for $d = 3$ focused on the basic question of whether or not a phase transition actually exists, but its existence is now well-established. Many random-field properties are solidly characterized and some aspects of the $d = 3$ phase transition remain unresolved, particularly the inconsistencies between the theoretical and experimental estimates of the universal critical parameters and the role of equilibration in experiments. The open questions provide fascinating opportunities to develop tools necessary to further explore this difficult area of statistical physics.

It quickly became evident in random-field experiments that different behaviors could be observed depending on the field and temperature histories of the crystals, something not often encountered in the pure and random-exchange systems. To describe the random-field experimental results, cooling and heating procedures need to be defined. Three of the most important protocols are listed in Table 4.11 and shown in Fig. 4.45. Cooling the dilute antiferromagnetic crystals in zero field, raising the field

Table 4.11: The most common of the experimental field-temperature protocols for experiments discussed in the text.

ZFC	cool to low T with $H = 0$, increase field to H, and heat
FC	cool in the field H
FH	after FC, heat with H still applied

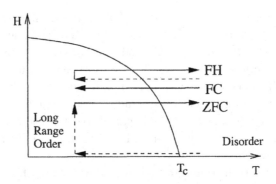

Fig. 4.45: Illustrations of three common experimental protocols for collecting data as described in Table 4.11.

and then increasing the temperature while taking data is known as zero-field cooling (ZFC). If, instead, the crystal is cooled in the field as data are taken, the procedure is known as field cooling (FC). After FC to low temperatures, data can be taken while reheating the sample, which is known as field-heating (FH). Each of these protocols can lead to very different experimental outcomes, which is one reason critical behavior characterizations of the $d = 3$ transition are so challenging and interesting.

4.7.1 $d = 2$ *Random-field Ising critical behavior*

For $d = 2$, the transition should be destroyed when random fields are introduced, for example, by the application of a uniform field to a dilute $d = 2$ antiferromagnet. Optical birefringence experiments (Ferreira *et al.*, 1983) on $Rb_2Co_{0.85}Mg_{0.15}F_4$ clearly show the destruction of the transition; Fig. 4.46 demonstrates that a small field $H = 20kOe$ is sufficient to

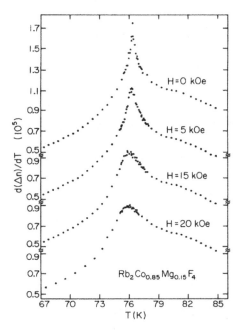

Fig. 4.46: The magnetic specific heat of the magnetically dilute antiferromagnet $Rb_2Co_{0.85}Mg_{0.15}F_4$ measured using $d(\Delta n)/dT$ for various applied fields (Ferreira *et al.*, 1983). The transition is quickly rounded by the random field and the peak is rounded independent of the field-temperature cycling, so the rounding is equilibrium behavior. This demonstrates that the $d = 2$ Ising phase transition is destroyed by the random fields, as expected. Reproduced with permission from American Physical Society.

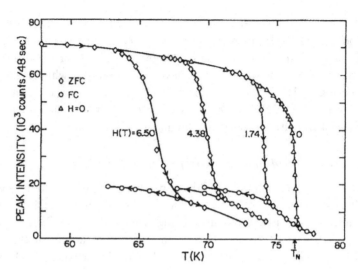

Fig. 4.47: The FC-ZFC hysteresis of the peak intensity versus T in $Rb_2Co_{0.85}Mg_{0.15}F_4$ for several applied fields (Belanger *et al.*, 1985c). The temperature $T_F(H)$ is defined to be the temperature of the steepest slope of the ZFC curves.

severely round the transition that appears sharp for $H = 0$. The behavior near the peak is independent of the field-temperature cycling, so it can be considered to be equilibrium behavior. FC-ZFC hysteresis is apparent at lower temperatures in neutron scattering data, as demonstrated in Fig. 4.47, where the ZFC and FC show different behavior in the peak Bragg scattering intensity, $I(H)$, well below the rounded peak (Belanger *et al.*, 1985c). A freezing temperature $T_F(H)$ can be defined by the temperature of the greatest slope, $-dI(H)/dT$, in the ZFC data. The shift of this temperature from the $H = 0$ transition temperature, $T_N - T_F(H)$, scales as $H_{rf}^{2/\phi_{rf}}$, as shown by the log–log plot in Fig. 4.48. A fit to the behavior yields the $d = 2$ random-field crossover exponent $\phi_{rf} = 1.74(2)$, which compares well to the value of γ (see Table 2.1).

Time dependence of the Bragg scattering at the temperature of the steepest slope can be observed (Belanger *et al.*, 1985c) by cooling in zero field to $T = 66.1$ K and then raising the field quickly to $H = 6.5$ T. The intensity versus the logarithm of time is shown in Fig. 4.49 along with a straight line fit.

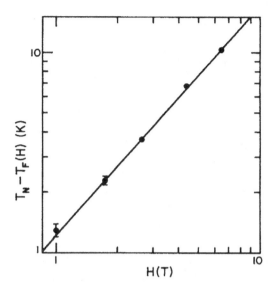

Fig. 4.48: A log–log plot of $T_N - T_F(H)$ versus H. The straight line is a fit yielding the crossover exponent $\phi_{rf} = 1.74(2)$ (Belanger *et al.*, 1985c).

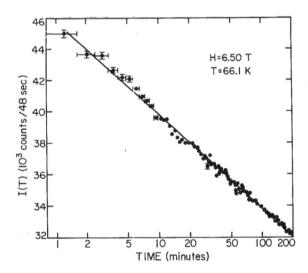

Fig. 4.49: A semi-log plot of the Bragg scattering intensity versus time for $Fe_2Co_{0.85}Mg_{0.15}F_4$ after cooling in zero field to $T = 66.1$ K and then raising the field to $H = 6.5$ T (Belanger *et al.*, 1985c).

4.7.2 $d = 3$ *Random-field Ising critical behavior*

The strength of the effective random field for a dilute antiferromagnet in a field is given (Fishman and Aharony, 1979) by

$$h_{\rm rf}^2 = \frac{x(1-x)(T_N^{\rm MF}(1)/T)^2(g\mu_B SH/k_B T)^2}{(1+\Theta^{MF}(x)/T)^2}, \tag{4.22}$$

where $T_N^{\rm MF}(1)$ is the mean-field value of the transition for the pure system, and $\Theta^{\rm MF}$ is the mean-field Curie–Weiss parameter. New critical behavior is expected for

$$|t| < h_{\rm rf}^{2/\phi_{\rm rf}}, \tag{4.23}$$

where $\phi_{\rm rf}$ is the random-field crossover exponent and $t = T - T_N + bH^2/T_N$, where b describes the mean-field temperature shift of the transition temperature given by $T_N^{\rm MF} = T_N - bH^2$. The free energy is predicted to have the scaling form

$$F = |t|^{2-\alpha} f(th_{\rm rf}^{-2/\phi_{\rm rf}}), \tag{4.24}$$

where $|t|^{2-\alpha}$ is the $H = 0$ critical behavior and $f(th_{\rm rf}^{-2/\phi_{\rm rf}})$ describes the crossover to the random-field behavior for $H \neq 0$. For new critical behavior, such as that observed for the $d = 3$ random-field Ising model, the new critical behavior can be derived from the equations above. For example, the magnetic specific heat can be written (Belanger and Young, 1991; Ferreira *et al.*, 1991c) in the form

$$C \propto |t|^{-\alpha} f(th_{\rm rf}^{-2/\phi_{\rm rf}}), \tag{4.25}$$

which can then be rewritten in the form

$$C \propto h_{\rm rf}^{-2\alpha/\phi_{\rm rf}} f(th_{\rm rf}^{-2/\phi_{\rm rf}}). \tag{4.26}$$

Close to the new phase transition, the function $f(th_{\rm rf}^{-2/\phi_{\rm rf}})$ should take the form $h^{2\alpha_{\rm rf}/\phi_{\rm rf}}|t - t_c|^{-\alpha_{\rm rf}}$, where $\alpha_{\rm rf}$ is the random-field specific heat exponent, $t_c = (T_c(H) - T_N + bH^2)/T_N = (ch_{\rm rf}^2)^{1/\phi_{\rm rf}}$, and $t - t_c$ is the reduced temperature shift from the new transition temperature. This gives

$$C \propto h_{\rm rf}^{2(\alpha_{\rm rf}-\alpha)/\phi_{\rm rf}}|t - t_c|^{-\alpha_{\rm rf}}. \tag{4.27}$$

Normally, we can simply use the usual power law form for the specific heat

$$C = A^{\pm}|t|^{-\alpha_{rf}} + B, \qquad (4.28)$$

using the new reduced temperature $t = (T - T_C(H))/T_C(H)$ to fit the data. The other new critical behaviors for the random-field transition can be derived in a similar manner.

Based on the ideas of universality, there was no expectation in early experiments that random-field effects would be influenced by the value of the magnetic concentration x other than determining the strength of the effective random field strength h_{rf}. As will be shown below, it appears that this is approximately the case for many of the universal critical exponents and amplitude ratios, but a large effect can be observed for data obtained from X-ray and neutron scattering experiments. As the transition is approached and $|t|$ decreases, the system exhibits a crossover from random-exchange to random-field critical behavior. The width of the asymptotic random-field critical region grows with the strength of the random field. To obtain data with reasonably large ranges of $|t|$ over which random-field critical behavior can be analyzed with easily accessible field strengths, typically $H \leq 7T$, significant dilution of the magnetic lattice is convenient. What was not realized is that the tendency to form metastable domains upon FC is concentration dependent. As will be discussed in Section 4.7.4, vacancies form a percolating lattice for concentrations $x < x_v$, where x_v is the vacancy percolation threshold and is equal to $1 - x_p$, where x_p is the magnetic site percolation threshold below which magnetic order can not occur for the dominant nearest-neighbor interactions. For the body-centered tetragonal lattices of $Fe_xZn_{1-x}F_2$ and $Mn_xZn_{1-x}F_2$, $x_v = 0.754$. As will be explained, the formation of FC domains for $x < x_v$ adds a level of complexity to the $d = 3$ experimental results for temperatures below the transition that is absent at higher concentrations. Nevertheless, numerous early experiments were done with $x < x_v$, and much was learned about the physics of random-fields and domain physics from them.

Early birefringence experiments on the highly anisotropic $Fe_xZn_{1-x}F_2$ and weakly anisotropic $Mn_xZn_{1-x}F_2$ systems (Belanger *et al.*, 1982b,c, 1983a) provided compelling evidence that the random-fields alter the shape of the specific heat peak for $d = 3$ dilute anisotropic antiferromagnets. In contrast to $d = 2$ systems, where the transition is clearly destroyed upon the introduction of random-fields, the $d = 3$ random-field specific heat behavior suggests an underlying phase transition associated with a universality class different from the zero field random-exchange case. That result, at a time when the existence of the new universality class was in question, added to the motivation for numerous additional experiments, theoretical studies, and computer simulations.

At about the same time, neutron scattering experiments with $x < x_v$ on the highly anisotropic $Fe_xZn_{1-x}F_2$ and $Co_xZn_{1-x}F_2$ and the weakly anisotropic $Mn_xZn_{1-x}F_2$ systems (Birgeneau *et al.*, 1985, 1983b; Cowley *et al.*, 1984b,c; Hagen *et al.*, 1983; Yoshizawa *et al.*, 1982) demonstrated highly non-Lorentzian line shapes and the inability of $d = 3$ systems to gain long-range antiferromagnetic long-range order upon FC, even at low temperatures. Although that had been interpreted as evidence supporting a destroyed transition for $d = 3$, as predicted by much of the theoretical work of the time, it was shown in later neutron scattering experiments (Belanger *et al.*, 1985b; Yoshizawa *et al.*, 1985) that long-range order appeared to be established using the ZFC procedure and persisted upon heating for temperatures $T < T_C(H)$. It was unknown from the early experiments whether the FC domain state or long-range order represented the low T equilibrium state, though the ZFC procedure resulted in scattering line shapes that were more consistent with the existence of a phase transition.

Neutron scattering studies for $T > T_C(H)$ explored (Belanger *et al.*, 1985b) the critical behaviors of the line shapes obtained for $Fe_{0.6}Zn_{0.4}F_2$. Figure 4.50 shows the difference between a scan with $H = 0$, which is consistent with a Lorentzian-like line shape, and one with $H = 2T$, which shows contributions from a Lorentzian-like and a squared-Lorentzian-like line shape. Note that, for $H = 2$ T, the scattering is dominated by the squared-Lorentzian contribution for small $|q|$. Figure 4.51 shows the half widths of scans taken upon ZFC, FC and FH with an applied field $H = 5.5$ T. The ZFC scans are resolution limited below $T_C(H)$. FC widths are larger than resolution, indicating domain formation. FH from low temperatures resulted in scans with more narrow widths than FC, but they are not resolution-limited. For $T > T_{eq}(H)$, the system is in equilibrium; the widths are all the same no matter what field-temperature cycling is done.

Fits of the line shapes for $T > T_{eq}(H)$ led to the conclusion that the measured random-field critical exponents and amplitude ratios associated with κ and χ more closely resemble those observed in $d = 2$ pure systems than $d = 3$ pure systems, albeit with significant uncertainties in the fitted parameters. This is complemented by the specific heat for $H > 0$, which appears to be close to a symmetric logarithmic behavior as would be expected in the pure $d = 2$ Ising model. Figure 4.52 shows the example of κ versus T at different values of H for $T > T_{eq}(H)$, where $T_{eq}(H)$ is the

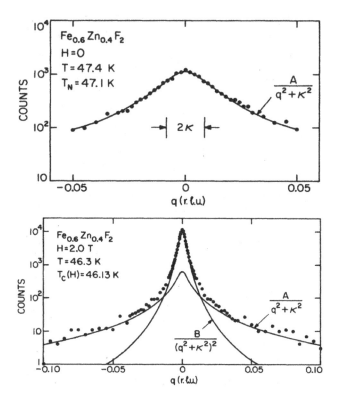

Fig. 4.50: The intensity versus q scans showing typical neutron scattering line shapes for $Fe_{0.6}Zn_{0.4}F_2$ for $H = 0$ (upper figure) and $H = 2$ T (lower figure). Note the different range of q in the two figures. The $H = 0$ line shape, at a temperature 0.3 K above T_N, is nearly Lorentzian, while the $H = 2$ T data, at a temperature 0.17 K above $T_C(H)$, are better described by a sum of a Lorentzian and squared-Lorentzian line shape. The squared-Lorentzian contribution dominates at small $|q|$ for $H = 2$ T (Belanger *et al.*, 1985b).

equilibrium boundary. It was also concluded, from the scattering results and the specific heat results, that the lower critical dimension, d_l, below which there would be no long range order, must satisfy the relationship $d_l < 3$. This contrasted with early theoretical work suggesting $d_l = 3$ for random-field Ising systems. More accurate scattering results for $x > x_v$ will be discussed later that suggest exponents close to, but different from the $d = 2$ pure Ising model exponents.

The interpretation of the scattering data for $x < x_v$ is greatly complicated by the stark difference between ZFC and FC line shapes for $T < T_{eq}(H)$, where $T_{eq}(H)$ lies just above $T_C(H)$, and questions regarding

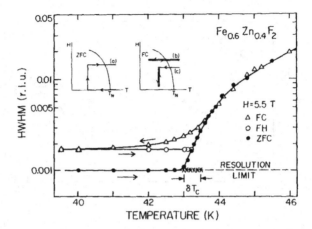

Fig. 4.51: The logarithm of the width of the neutron scattering line shapes for $Fe_{0.6}Zn_{0.4}F_2$ with $H = 5.5$ T as a function of T. ZFC scans show resolution limited Bragg peaks below the transition. FC scans are wider than resolution and wider than ZFC for $T < T_{eq}$. FH scans are more narrow than FC scans near $T_C(H)$, but are still wider than resolution. The data also show that the width can decrease upon cooling, but does not increase upon heating. The expected spread in $T_C(H)$ is shown as δT_C and is a result of the measured concentration variation (Belanger *et al.*, 1985d).

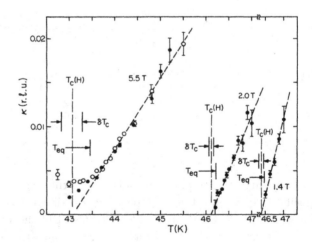

Fig. 4.52: The values of κ obtained (Belanger *et al.*, 1985b) from fits to the Lorentzian plus squared-Lorentzian contributions to the scattering line shapes for $Fe_{0.6}Zn_{0.4}F_2$ versus T for different H (Belanger *et al.*, 1985b). The straight lines show fits to a linear behavior with ν set to unity. The expected rounding, given by δT, grows with H because of the dependence of the random-field dependence on x given in Eq. (4.22).

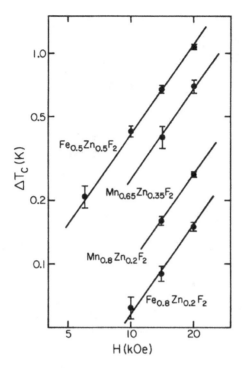

Fig. 4.53:   The $d = 3$ random-field Ising crossover exponent $\phi_{rf} = 1.4$ measured in $Fe_x Zn_{1-x} F_2$ and $Mn_x Zn_{1-x} F_2$ for concentrations $x < x_v$ and $x > x_v$ (Belanger *et al.*, 1982c). A similar, but more comprehensive list, can be found in (Ferreira *et al.*, 1991c). An accepted value for the crossover exponent is $\phi_{rf} = 1.42(3)$.

the correct equilibrium ground state. The $x = 0.6$ crystal also suffers from a significant variation in concentration that leads to a variation of $T_C(H)$ throughout the sample, as indicated in Fig. 4.52.

The crossover exponent, ϕ_{rf}, which governs the field dependence of the phase transition, appears to be robust. Figure 4.53 shows $T_C(H) - T_N$, determined from fits to the specific heat behavior (Belanger *et al.*, 1982c) versus H for different concentrations of $Fe_x Zn_{1-x} F_2$ and $Mn_x Zn_{1-x} F_2$. The scaling behavior, indicated by the straight lines is governed by the random-field crossover exponent ϕ_{rf}. The exponent is not sensitive to the concentration x or to the strength of the anisotropy of the system. The figure includes data for $x < x_v$ and $x > x_v$. It also includes data obtained in the strongly $(Fe_x Zn_{1-x} F_2)$ and weakly $(Mn_x Zn_{1-x} F_2)$ anisotropic systems. The value $\phi_{rf} = 1.4$ was used in the figure, but more recent work has established the value $\phi_{rf} = 1.42(3)$.

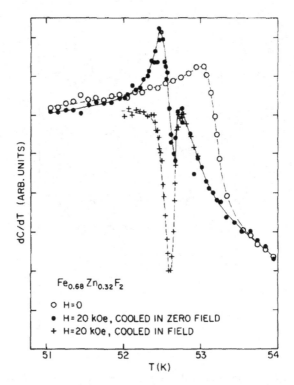

Fig. 4.54: The behavior of the capacitance of $Fe_{0.68}Zn_{0.32}F_2$ as a function of T for $H = 0$ and, for ZFC and FC, $H = 20kOe$ (Rezende *et al.*, 1984). The complicated behavior for $H \neq 0$ is discussed in the text. The point below which FC and ZFC differ, $T_{eq}(H)$, is extremely well-defined. Reproduced with permission from AIP Publishing.

The capacitance technique is well-suited for the study of the scaling behavior of $T_{eq}(H)$ (King *et al.*, 1985). The expansion of the crystal and its dielectric properties contribute to the capacitance signal. For $H = 0$, the combined effect, shown in Fig. 4.54 for $Fe_{0.68}Zn_{0.32}F_2$, is to reproduce the specific heat behavior, just as it does for FeF_2 (see Section 3.4). However, for $H \neq 0$, the behavior is starkly different for the ZFC and FC procedures. This is convenient for determining the equilibrium boundary $T_{eq}(H)$ to be the temperature where the ZFC and FC behaviors sharply diverge from one another.

Just as with phase boundary $T_C(H)$, the equilibrium boundary scales with the random-field variable $H^{2/\phi_{rf}}$, as shown in Fig. 4.55 for two widely separated concentrations, $x = 0.72$ and 0.46, both below x_v. The difficulties with FC domain formation at low temperatures have been, in part,

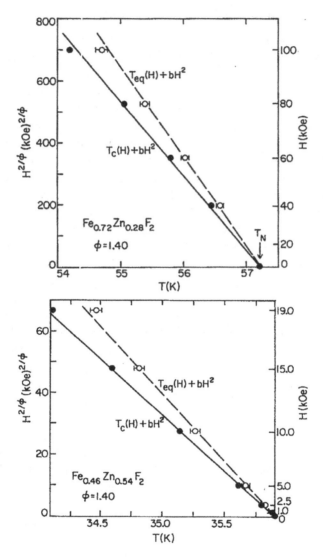

Fig. 4.55: The scaling of $T_C(H)$ and $T_{eq}(H)$ for $Fe_{0.72}Zn_{0.28}F_2$ (upper figure) and $Fe_{0.46}Zn_{0.54}F_2$ (lower figure) (King *et al.*, 1985).

addressed by experiments with $x > x_v$, to be described below. Difficulties in correctly analyzing the scattering line shapes near $T_C(H)$, however, remain for $x > x_v$, and that is one outstanding problem yet to be resolved for the $d = 3$ random-field Ising model, as will be discussed further below.

4.7.3 $d = 3$ *Random-field dynamics for* $x < x_v$

The dynamics of the random-field Ising model are so slow that they affect experiments that are normally considered to yield static critical behavior. This has led to much of the difficulty in the interpretation of the data. There are effects from the extremely slow dynamics for both FC and ZFC in $Fe_xZn_{1-x}F_2$ for $x < x_v$. This is visible in the long-time evolution and the ac-susceptibility. While the order-parameter measurements reveal long relaxations in FC measurements for $x < x_v$, measurements for $x > x_v$, as discussed in Section 4.7.7, show little or no long-time dependence on laboratory time scales. Nevertheless, the system for $x > x_v$ shows metastability below a temperature $T_{eq}(H)$ just above $T_C(H)$. Hysteresis is observed for $x > x_v$ with respect to temperature, but time dependence is not observed.

For $x < x_v$, the spin echo neutron scattering technique showed no dynamic critical behavior for $H > 0$; on the time scales involved, the dynamical behavior is just too slow. Time dependent behavior on laboratory time scales was observed using neutron scattering (Belanger *et al.*, 1987). Figure 4.56 shows the scattering line shapes for the $Fe_{0.46}Zn_{0.54}F_2$ sample (Belanger *et al.*, 1987), which has an extraordinarily small variation in concentration, as discussed in Section 3.7, and a corresponding spread in T_C of 0.007 K at $H = 0$. Figure 4.57 shows the behavior of the ZFC and FC scattering intensity observed in $Fe_{0.46}Zn_{0.54}F_2$ at the Bragg peak for $H = 1.5$, 1.9, and 3 T and just off the Bragg peak for $H = 1.9$ T. The transition temperature $T_C(H)$ and equilibrium temperature $T_{eq}(H)$ have been determined by capacitance measurements (King *et al.*, 1985) and $T_{eq}(H)$ in the neutron and capacitance measurements are consistent. The ZFC scattering intensity is limited by the severe extinction in the well-ordered ZFC state compared to the FC domain state. The ZFC intensity rises as T approaches $T_C(H)$ from below as the long-range ZFC order breaks down as extinction is relieved. Figure 4.58 shows the time evolution of the scattering intensity at different fields and temperatures after cooling in zero field to a temperature close to $T_C(H)$ and then raising the field. The time dependence was measured over three decades in time, with a minimum at 30 seconds.

Time dependence can also be observed using the capacitance technique (Belanger *et al.*, 1985d). Upon FC, the time dependence of the capacitance was measured over two decades of time in $Fe_{0.68}Zn_{0.32}F_2$ for times as large as 20 minutes, as shown in Fig. 4.59. Clearly, long-time dynamics are present for $x < x_v$.

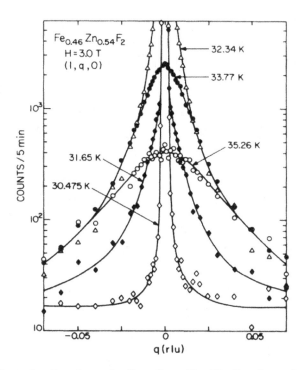

Fig. 4.56: Scattering line shapes for $Fe_{0.46}Zn_{0.54}F_2$. The line shapes for $T < T_C = 32.115$ K were taken after ZFC with $H = 3$ T. The data for $T > T_C$ are adequately fit with modified Lorentzian plus squared-Lorentzian line shapes, but the data for $T < T_C$ are not (Belanger *et al.*, 1987).

The activated dynamics associated with the random-field Ising model is difficult to characterize, but that was accomplished in a high quality sample with $x < x_v$. The ac-susceptibility was studied (King *et al.*, 1986) in the $Fe_{0.46}Zn_{0.54}F_2$ that has a very small variation of the magnetic concentration. Figure 4.60 shows $\chi'(\omega)$ versus T for $H = 0$ and $H = 1$ T and, in the insert, ZFC and FC measurements at $H = 2$ T are compared, with no apparent hysteresis. The lack of hysteresis shows that the real part, $\chi'(\omega)$, does not reflect large scale domain wall dynamics. Figure 4.61 shows the frequency dependence at $H = 1$ T. Lower frequency yields sharper peaks, indicating the rounding is a consequence of the random-field dynamics. The ac-susceptibility peaks are much more rounded than the specific heat peak, again demonstrating that this is a dynamic effect. Faraday rotation studies (Kleemann *et al.*, 1986), which are nominally static measurements,

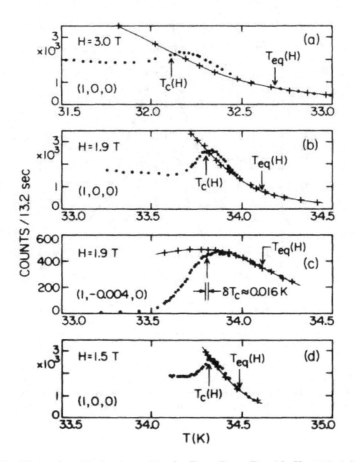

Fig. 4.57: The peak scattering intensities for $Fe_{0.46}Zn_{0.54}F_2$ with $H = 1.5$, 1.9 and 3 T, and the scattering for $q = 0.004$ for $H = 1.9$ T using the FC and ZFC protocols (Belanger *et al.*, 1987). The intensities for ZFC are more severely limited by extinction effects.

nevertheless exhibited rounding attributable to dynamic effects. The range of frequency was greatly extended in the measurements (Nash *et al.*, 1991) so that it could be concluded that conventional dynamics do not hold, but activated dynamics, as proposed by Villain (1985), Fisher (1986) and Ogielski and Huse (1986) works well. The background-corrected $\chi'(\omega)$ versus the logarithm of $|t|$ is shown in Fig. 4.62. The data saturate at higher values for lower frequencies as $|t|$ decreases, but the frequency dependence is slow. After correcting for the background, the peak value of $\chi'(\omega)$ versus the logarithm of the frequency over nearly eight decades is shown in Fig. 4.63.

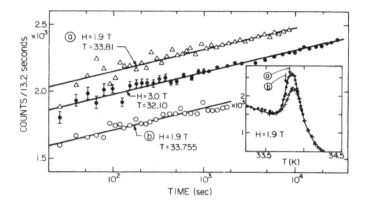

Fig. 4.58: Scattering intensities versus time at the antiferromagnetic Bragg point for $Fe_{0.46}Zn_{0.54}F_2$ after cooling in ZFC and then raising the field close to $T_C(H)$ (Belanger *et al.*, 1985d). The inset shows the temperatures for $H = 1.9T$. The time dependence is approximately logarithmic for all three cases.

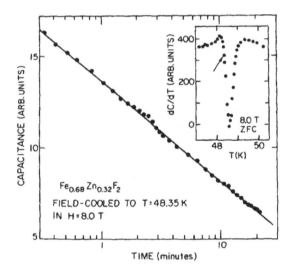

Fig. 4.59: The capacitance, C, versus time for $Fe_{0.68}Zn_{0.32}F_2$ after FC (Belanger *et al.*, 1985d). The data are taken at the temperature indicated in the inset, which shows dC/dT versus T after FC.

The curve is a fit to activated dynamics. If normal dynamic behavior held, the good fit should be to a straight line. The activated dynamics proposal is based on the domination of random-field, rather than thermal, fluctuations near the phase transition. It was proposed that the characteristic time for

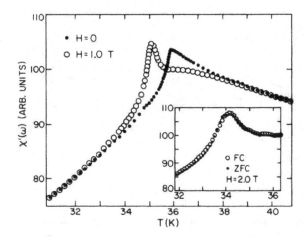

Fig. 4.60: $\chi'(\omega)$ versus T in $Fe_{0.46}Zn_{0.54}F_2$ for $H = 0$ and $H = 1$ T. In the field, a sharp peak is expected near the transition, whereas for $H = 0$, it is a cusp (King *et al.*, 1986). Reproduced with permission from American Physical Society.

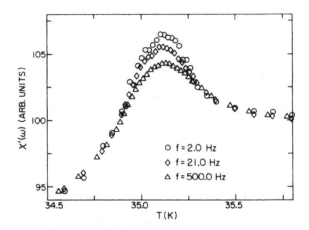

Fig. 4.61: $\chi'(\omega)$ versus T for three frequencies in $Fe_{0.46}Zn_{0.54}F_2$ with $H = 1$ T. The lowest frequency shows the sharpest peak (King *et al.*, 1986). Reproduced with permission from American Physical Society.

critical fluctuations scales as $\tau \propto \exp \xi^\theta$ rather than the usual scaling $\tau \propto \xi^\theta$, where $d-\theta = (2-\alpha)/\nu$ and θ is an effective dimensionality. Further work on the dynamics of random systems, including random-field systems has been done, for example in Shapir (1985, 1987) and Nattermann *et al.* (1990). Note that the ac-susceptibility measurements were done using a crystal

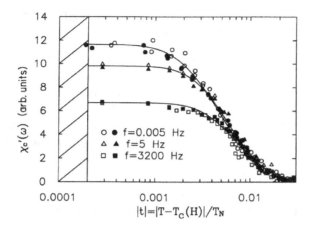

Fig. 4.62: $\chi'(\omega)$ versus the logarithm of $|t|$ i over a wide range of frequency, approximately six decades measured in the Fe$_{0.46}$Zn$_{0.54}$F$_2$ sample that has a very small variation of x which would affect the results only within the shaded region. The open and filled points are for $T < T_C(H)$ and $T > T_C(H)$, respectively (Nash *et al.*, 1991). Reproduced with permission from American Physical Society.

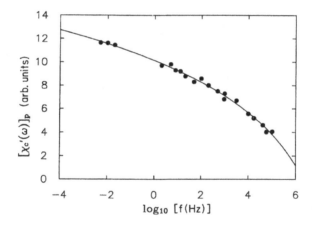

Fig. 4.63: The peak value of $\chi'(\omega)$ after correcting for the background versus the logarithm of the frequency. The fit is to activated dynamics. Conventional dynamics would be represented by a straight line (Nash *et al.*, 1991). Reproduced with permission from American Physical Society.

with $x < x_v$. It would be fascinating to measure the ac-susceptibility for $x > x_v$ to see how it might differ from $x < x_v$. The weakening of the magnetic lattice by the percolating vacancy lattice might significantly amplify the dynamic effects.

4.7.4   *Vacancy percolation and the stability of $d = 3$ antiferromagnetic long-range order*

The observation in the early studies of new critical behavior and the inability to achieve long-range order below the transition temperature using FC procedures motivated a Monte Carlo study (Barber and Belanger, 2000) to understand why low-T FC domains easily form upon FC. To explore the stability of the antiferromagnetic long-range order, the Monte Carlo simulations were performed for $d = 3$ body-centered antiferromagnetic lattices with various concentrations of magnetic vacancies. The exchange interaction energies were chosen to simulate the temperature behavior of $Fe_x Zn_{1-x} F_2$. The staggered magnetic moment, shown in Fig. 4.64, illustrates the typical difference between ZFC and FC data that can occur just below the transition. The behavior evolves as x varies. The concentrations $x = 0.60$ (a) and 0.70 (b) show strong differences between ZFC and FC data. The difference at $x = 0.70$ is small. For $x = 0.80$ (c) and

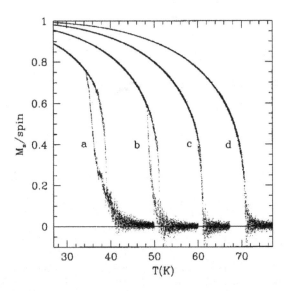

Fig. 4.64: Monte Carlo simulations (Barber and Belanger, 2000) of the staggered magnetization of a diluted Ising antiferromagnet on a body-centered tetragonal lattice with exchange interactions that approximately mimics the behavior of $Fe_x Zn_{1-x} F_2$ for $x = 0.60$ (a), 0.70 (b), 0.80 (c), and 0.90 (d). with $H = 13$ T. The simulations of the staggered magnetization using the ZFC and FC show different behavior below T_C for a and b, but not c or d. Simulations with finer steps show that difference between ZFC and FC vanishes close to $x_v = 0.754$, the site percolation threshold for vacancies.

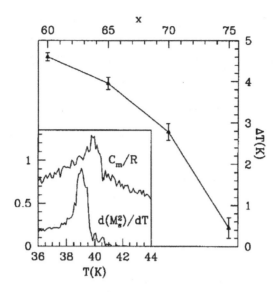

Fig. 4.65: Stability of the antiferromagnetic long-range order of $Fe_xZn_{1-x}F_2$. The maximum separation between the FC and ZFC staggered magnetizations is plotted versus x and approaches zero near the vacancy percolation threshold at $x_v = 0.754$ (Barber and Belanger, 2000). In the simulation, the specific heat and temperature derivative of the order parameter were calculated and these are shown in the inset. The two peaks should not coincide (Belanger *et al.*, 1996); it was suggested that they do (Birgeneau *et al.*, 1996; Hill *et al.*, 1997).

0.90 (d), there is no discernible difference. Figure 4.65 shows that the difference between ZFC and FC vanishes at a concentration near $x_v = 0.754$, which is the percolation threshold concentration for the magnetic vacancies for a body-centered tetragonal lattice. The scattering line shapes at $x = 0.76$ show (Barber *et al.*, 2004) evidence for scattering from the fractal structure of the vacancies.

For $x > x_v$, the simulation data are independent of the field and temperature cycling procedure used and long-range order is observed for all $T < T_C$. It is not clear that the concentration above which stability against FC domain formation upon FC that persists to low temperatures coincides precisely with x_v. However, for $x > x_v$, vacancies only form isolated clusters, whereas for $x < x_v$, the vacancy network percolates throughout the lattice and this may weaken ability to achieve long-range magnetic order. Percolation properties evolve quickly as a function of x, so it is reasonable to expect that the weakening of the magnetic lattice becomes significant for x less than but very close to x_v.

Up to the time that the significance of x_v was recognized, most experiments had been done with $x \leq 0.72$. For the $d = 3$ body-centered lattice, $x_v = 0.754 = 1 - x_p$, where x_p is the magnetic percolation threshold for nearest-neighbor interactions and is the concentration below which no magnetic long-range order can take place. For the concentration range $x_p < x < x_v$, both the magnetic and vacancy lattices percolate throughout the lattice. As an aside, for the $d = 2$ square lattice, $x_c = 0.593$ and $x_v = 0.437$, so there is no region where the $d = 2$ systems can have long-range order coexisting with a percolating vacancy lattice; the destruction of long-range order for $d = 2$ is purely a random-field effect that is not influenced by the vacancy percolation effect and is consistent with the equilibrium behavior.

The magnetic lattice for $x = 0.84$ is stable to relatively large fields (Sakon *et al.*, 2002). Figure 4.66 shows susceptibility data taken using two different magnets. The upper panel shows data taken up to fields $H = 13.5$ T that scale with the crossover exponent $\phi = 1.42$. Data for fields $12 \leq H \leq 20$ T, taken on a separate high-field magnet, are more sparse but, for $H \leq 18$ T, the scaling behavior continues to be followed. For 20 T, the data clearly deviate from the scaling behavior seen at lower fields and this shows the instability of the antiferromagnetic long-range order. This field is much larger than would be anticipated by extrapolating the field above which glassy behavior is observed as it varies with x (Barber *et al.*, 2004; Montenegro *et al.*, 1998, 1999, 2000c; Rosales-Rivera *et al.*, 2000), as shown in Fig. 4.67.

It is important to note that x_v plays less of a role in the $H = 0$ random-exchange Ising model experiments, though the measured dynamic critical behavior might be affected, as mentioned in Section 4.6.5. In that case, there is no field frustrating the formation of long-range order, so long-range antiferromagnetic correlations grow without impediment, though slowly.

With random-fields introduced, x_v separates the region $x > x_v$ where antiferromagnetic long-range order will be achieved at low T regardless of the temperature and field cycling process, and the region $x < x_v$ where domain structure easily forms and persists to low T. Domain structure will greatly affect the scattering line shapes and can obscure the ordering process. This does not, in of itself, necessarily mean there is no long-range order. Although there is not one long-range antiferromagnetic lattice, there is nothing to prevent two competing long-range antiferromagnetic lattices to exist. However, scattering techniques have difficulty distinguishing true short-range order from two fractal-like long-range

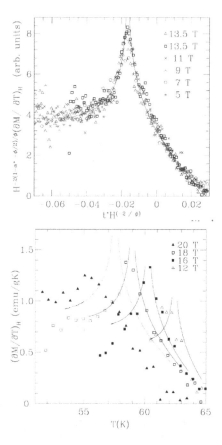

Fig. 4.66: The behavior of $d\chi/dT$ for $Fe_{0.84}Zn_{0.16}F_2$ (Sakon *et al.*, 2002). The upper panel shows the scaling behavior for fields $H \leq 13.5$ T using the crossover exponent $\phi_{rf} = 1.42$. The lower panel shows $(\partial M/\partial T)_H$ versus T. The curves represent the same scaling function as used in the upper panel. The data follow those curves until $H = 18$ T. However, the scaling fails at $H = 20$ T, indicating that the antiferromagnetic long-range order breaks down at a field between 18 and 20 T.

clusters that are intertwined. Hence, the observations of domain structures and well-behaved critical behavior are not necessarily mutually exclusive. On the other hand, the rounding of the phase transition upon FC suggests that correlations in thermal fluctuations are limited in size. Furthermore, considering the interface energy cost of having two infinite, but interfacing domains, the ground state is likely that with one antiferromagnetically ordered lattice. Although $x > x_v$ allows a more straightforward interpretation of scattering experiments, there are two things to keep in mind. First,

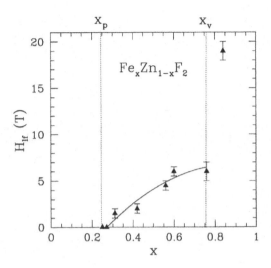

Fig. 4.67: Stability of the antiferromagnetic long-range order of $Fe_xZn_{1-x}F_2$ (Sakon *et al.*, 2002). The highest field where a transition from paramagnetism to antiferromagnetic order is shown versus x. The extrapolation of the values below x_v is far below the value for $x = 0.84$, showing the greater stability for $x > x_v$. The stability points for $x < x_v$ can be found from several experimental works (Sakon *et al.*, 2002).

the critical parameters that can be observed for $x < x_v$ and $x > x_v$ are consistent even though scattering results are largely only from $T > T_{eq}(H)$ for $x < x_v$. Second, hysteresis is still observed extremely close to T_C and at the largest length scales involved in Bragg scattering for $x > x_v$. The most important advantage to having $x > x_v$ is to limit the hysteresis in the neutron scattering line shapes for $|q| > 0$, especially for $T < T_C$.

Having established that $x > x_v$ ensures stability against domain structure at low T, random-field experiments used to characterize the new universality class behavior were done using $Fe_xZn_{1-x}F_2$, with $x = 0.84$, 0.85 and $x = 0.93$. To probe the critical region at these high values of x requires collecting data with very small values of $|t|$ while applying large fields.

4.7.5 $d = 3$ Random-field order parameter

The most accurate determination of the order-parameter exponent for the random-field Ising system is from X-ray scattering measurements on $Fe_{0.85}Zn_{0.15}F_2$. This is a case where the concentration is above the vacancy percolation threshold $x_v = 0.76$. The values $\beta = 0.16(2)$ (Ye *et al.*, 2006)

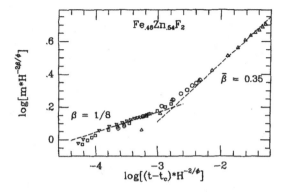

Fig. 4.68: The $d = 3$ Ising random-exchange and random-field order-parameter measured using anomalous dilation in $Fe_{0.46}Zn_{0.54}F_2$ for fields $H = 0.14$, 0.5, 1.5, and 3.0 T (Ramos *et al.*, 1988b). Reproduced with permission from EDP Sciences.

and $\beta = 0.17(1)$ (Ye *et al.*, 2002) were obtained. These measurements will be discussed in more detail below. One striking feature is the sharp crossover from random-exchange to random-field critical behavior. The sharp crossover feature was observed (Ramos *et al.*, 1988b) in earlier measurements of the anomalous dilation in $Fe_{0.46}Zn_{0.54}F_2$, for which $x < x_v$. The value $\beta \approx 1/8$ was obtained, as shown in Fig. 4.68. Few theoretical estimates for β are close to the measured values, as will be discussed below.

The order parameter for the $d = 3$ random-field Ising model for $x > x_v$ was measured (Ye *et al.*, 2006, 2002) using synchrotron scattering, a technique that avoids extinction effects that are encountered in the neutron scattering experiments, as explained in Section 4.7.2. The crystal used, $Fe_{0.85}Zn_{0.15}F_2$, has a magnetic concentration above the vacancy percolation threshold, $x_v = 0.754$, and the low T state is long-range antiferromagnetic order regardless of whether the staggered magnetization is measured in $H = 0$, or $H > 0$ using either the ZFC or FC procedures, so long-range order is clearly the ground state.

For $H = 0$, no hysteresis is observed in the X-ray data; the system remains in equilibrium even near the transition. Once the field is applied, the peak scattering intensity at $q = 0$ measured in the X-ray experiments exhibits strong hysteresis in the behavior of the the phase transition. The hysteresis is so severe that the temperature controller feedback must be set to overdamped conditions. In ZFC experiments, as the T set point is changed, the temperature cannot be allowed to overshoot its set point temperature by even 10^{-3} K or the data will not exhibit a smooth change

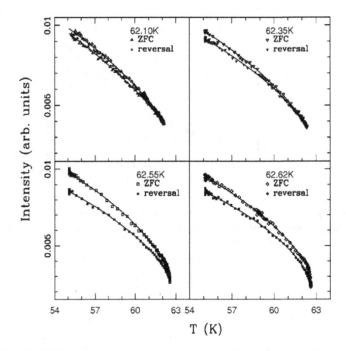

Fig. 4.69: The ZFC Bragg scattering intensity versus T of $Fe_{0.85}Zn_{0.15}F_2$ for $H = 10$ T and the intensity upon subsequent cooling after reaching temperatures shown in each panel. The transition temperature at this field is $T_C(H) = 62.85$ K.

with T. However, once the temperature stabilizes without an overshoot, no time dependence of the change in the ZFC intensity is observed in the experiments. The intensity upon subsequent cooling, however, does not achieve ZFC intensities.

The temperature hysteresis can be illustrated by temperature reversals at $H = 10$ T near to $T_C(H)$, as shown in Fig. 4.69. Approaching the transition, the ZFC data appear stable, but with a temperature reversal, the data fall below the ZFC data. The closer the reversal is to $T_C(H)$, the larger is the difference. The lower right panel is only 0.2 K below the transition and even the shapes of the two curves are different. As the transition is approached, the amount of order decreases, but the order does not increase as much once the temperature decreases. The effect is similar to the difference between FC and ZFC. Cooling cannot achieve the same amount of long-range order as heating after ZFC. On the other hand, both FC and ZFC show similar behavior if the intensity is integrated over a small range of $|q|$ around the Bragg point, as discussed below.

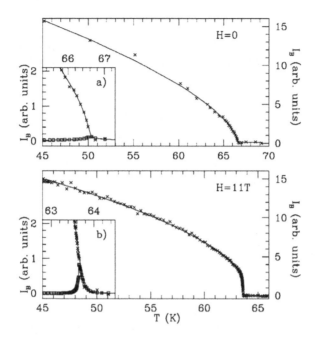

Fig. 4.70: The ZFC Bragg scattering versus T of $Fe_{0.85}Zn_{0.15}F_2$ for $H = 0$ and $H = 11$ T. The insets show the critical scattering contributions determined from neutron scattering, which are much larger for $H = 11$ T. The apparent Bragg scattering intensity does not change with time (Ye *et al.*, 2006).

Figure 4.70 shows the behavior of the staggered magnetization obtained (Ye *et al.*, 2006) using the ZFC procedure for $H = 0$ and 11 T. The main figures show the $q = 0$ scattering intensity, which represents the square of the staggered magnetization, $I_B(T) = M_S^2(T)\delta(q)$. The insets show the contributions to the intensity from fits to the critical scattering from fluctuations; they are small contributions relative to the Bragg scattering. The logarithm of the ZFC Bragg scattering intensity I_B versus the logarithm of T in Fig. 4.71 shows the random-exchange ($H = 0$) and random-field critical behaviors. For $H = 0$, the straight-line behavior shows the random-exchange power law, and a fit to the data yields with the fitted exponent $\beta = 0.35(2)$, With the application of the uniform field, a surprisingly sharp change in slope near $T_C(H)$ indicates the crossover to random-field behavior with an order-parameter exponent $\beta = 0.17(1)$. The crossover point scales with $H^{2/\phi_{rf}}$, as shown in Fig. 4.72.

The crossover from random-exchange to random-field critical behavior is observed for the ZFC data, but FC data for $q = 0$ do not exhibit the

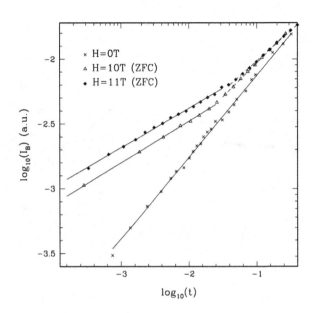

Fig. 4.71: A log–log plot of the ZFC Bragg intensity versus $|t|$ of $Fe_{0.85}Zn_{0.15}F_2$ with $H = 0$, 10 and 11 T. For the $H = 0$ random-exchange Ising order parameter, $\beta = 0.35(2)$. For $H > 0$, the order parameter critical behavior sharply crosses over from random-exchange to random-field Ising, with $\beta = 0.16(2)$ (Ye *et al.*, 2002).

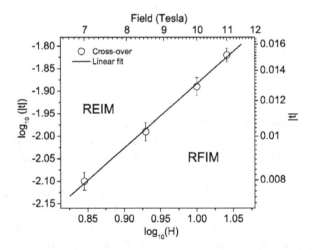

Fig. 4.72: The crossover from random-exchange to random-field Ising order-parameter critical behavior in $Fe_{0.85}Zn_{0.15}F_2$, which is governed by ϕ_{rf} (Ye *et al.*, 2006). The straight line fit represents $\phi_{rf} = 1.42$.

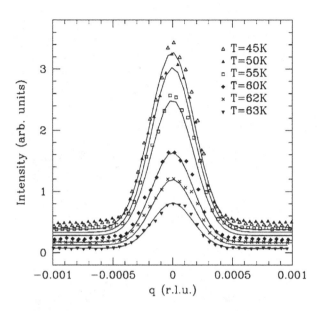

Fig. 4.73: ZFC resolution-limited $Fe_{0.85}Zn_{0.15}F_2$ line shapes below $T_C(H)$ for $H = 11$ T (Ye *et al.*, 2006).

crossover and the intensity remains below the ZFC intensity and nearly at the $H = 0$ levels for the entire range of $|t|$. The effect can be seen from the scans around $q = 0$ shown in Figs. 4.73 and 4.74. The ZFC line shapes are more narrow than FC or FH and the $q = 0$ intensity is higher than FC and FH. The $q = 0$ scattering within the instrumental resolution observed upon ZFC is spread out upon FC, so the $q = 0$ intensity misses some of the Bragg scattering intensity. However, the scattering integrated over a narrow range of q scales in the same way with the same intensity as shown in Fig. 4.75. The behavior of the integrated intensity demonstrates that the only difference between ZFC and FC is on the longest length scales over the entire crossover region. FC does not achieve long-range random-field Ising order, but the amount of order is equivalent if slightly smaller length scales are included. It is likely that the time needed to equilibrate order grows rapidly with the increase in the length scale of the order. It is the Bragg scattering from the longest length random-field-induced structure that cannot take place in FC, but the neutron scattering technique cannot reach the those length scales where hysteresis occurs. Hence, the neutron scattering data appear to be in equilibrium whether ZFC, FC, or FH protocols are followed.

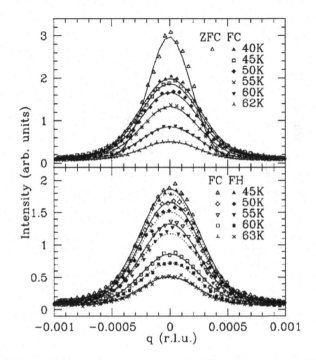

Fig. 4.74: FC $Fe_{0.85}Zn_{0.15}F_2$ line shapes below $T_C(H)$ compared to a ZFC line shape (upper panel) and a comparison of FC and FH line shapes below $T_C(H)$ (lower panel) (Ye *et al.*, 2006). The FC and FH line shapes are slightly wider than resolution and the peak values are smaller than ZFC peak values. Note that the length scales are greater than 1000 Å.

The reduction in intensity at $q = 0$ upon FC can be seen in Monte Carlo simulations (Shelton *et al.*, 2004), but in the simulations the intensity eventually jumps from the FC to the ZFC intensity well below $T_C(H)$. This suggests that the FC state, which lacks ordering on the longest length scales in the experiments, represents a metastable state.

The ZFC Bragg scattering data for $H > 0$ extend to reduced temperatures closer than 10^{-3}. This is comparable to the best critical order parameter studies in many pure systems.

Estimates for the value of β from numerical simulations tend to be very small, However, one recent simulation (Xiong and Xu, 2019) obtained the value $\beta = 0.173(2)$, in excellent agreement with the experimental value. On the other hand, the value for the specific heat exponent, $\alpha = 0.369(8)$, is far from the behavior observed in experiments, as discussed in the following section.

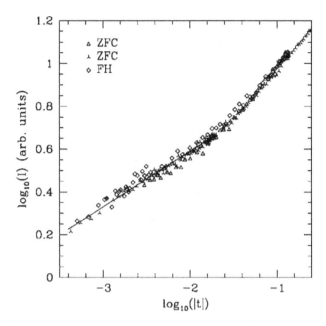

Fig. 4.75: The integrated intensities versus $|t|$ in a log–log plot for $Fe_{0.85}Zn_{0.15}F_2$ (Ye *et al.*, 2006). Although the FC and FH Bragg intensities fall below the ZFC intensities in the random-field region, the integrated intensities over a narrow region around the Bragg point all agree. This demonstrates that the intensity gets slightly spread out by the large FC domain structure.

4.7.6 $d = 3$ *Random-field specific heat*

Neutron and X-ray scattering experiments are studies of the structure of the magnetic system. For $q = 0$, it is the long-range order that is probed and for $q \neq 0$ it is the short-range structural fluctuations. The specific heat and $d(\Delta n)/dT$, on the other hand, are probes of thermal energy fluctuations that take place on that magnetic structure. They cannot probe long-range order except that the thermal fluctuations with long-range correlations can be suppressed if the structures on those length scales are missing. Only extremely close to the transition, where the correlation length for fluctuations, $\xi = 1/\kappa$, should be diverging, can a limitation on the size of fluctuations be seen in the form of rounding of the specific heat peak. This is similar to what is observed in simulations in the form of finite-size effects. Figure 4.76 shows the magnetic specific heat behavior $(d(\Delta n)/dT)$ versus T close to $T_C(H)$ for $x = 0.93$ and $H = 0$, 5, and 7 T. A small rounding occurs upon FC compared to ZFC data over 0.1 K around $T_C(H)$ for

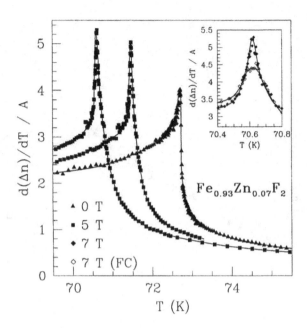

Fig. 4.76: The magnetic specific heat of $Fe_{0.93}Zn_{0.07}F_2$ (Slanič and Belanger, 1998).

$H = 7\,T$. This could be expected because, as seen from the X-ray scattering experiments, the random-field structure at the longest length scales is present with ZFC but not FC, so thermal fluctuations on that scale are suppressed in FC measurements.

Although hysteresis has been observed in the specific heat using the birefringence, as discussed above and in Ferreira *et al.* (1991c), and heat pulse techniques (Dow and Belanger, 1989), no hysteresis was observed using the relaxation technique where the sample is cooled during the measurement (Satooka *et al.*, 1998) even though the resolution and sharpness of the transition seem adequate to do so. The authors correctly comment that the technique used might obscure the hysteresis, which is subtle in the specific heat measurements. As shown in Section 4.4, where order-parameter measurements were discussed, cooling, even over small temperatures, can produce different results than techniques that involve only increasing temperatures and this might be another important manifestation of that effect. By heating to a temperature above the measurement temperature and then cooling to obtain the specific heat, some long-range fluctuations could be suppressed.

Other specific heat measurements (Birgeneau *et al.*, 1996; Hill *et al.*, 1997) also showed no hysteresis. Although the specific heat technique is not described in detail, other techniques described in the work do show hysteresis as well as rounding. It would be remarkable if hysteresis did not show some effect in the fluctuation contributing to the specific heat, some of which occur on long-length scales close to the transition. As discussed in Section 3.3.2, one possible explanation is that the specific heat sample may have enough concentration variation to obscure the hysteresis because of rounding and that effect increases with the applied field strength. The sample uniformity characterization was not provided, but the points near the specific heat peak are few in number. If the sample temperature was allowed to reach temperatures slightly above the measuring temperature, the possible effect described in the previous paragraph might be important. There are a number of possible reasons the FC-ZFC hysteresis was not observed. On the other hand, there is no convincing experimental evidence that the subtle hysteresis is absent from the random-field Ising specific heat near the phase transition.

The specific heat peak for $H = 0$ has the asymmetric shape of the random-exchange model with $\alpha < 0$, as discussed. Close to $T_C(H)$, the specific heat peak for $H > 0$ is symmetric and approximately logarithmic (Slanič and Belanger, 1998). The scaling behavior is demonstrated for $Fe_{0.93}Zn_{0.07}F_2$ in Fig. 4.77 where the vertical axis shows $Ch_{rf}^{2\alpha_{re}/\phi_{rf}}$ plotted versus $th_{rf}^{-2/\phi_{rf}}$. The data collapse nicely using the crossover exponent $\phi_{rf} = 1.42$. The rounding for small H is accentuated by the temperature scaling. The solid curves show the behavior of a symmetric logarithmic peak shape that describes the data well.

The critical behavior of the specific heat has been measured since some of the earliest experiments and the results have always indicated an exponent close to zero and an amplitude ratio close to unity (the precise value could not be obtained from these data), regardless of the concentration x. On the simulation side, values for α have been obtained that range from large and negative to large and positive, without much consistency. The difficulty in determining the exponent from simulations has been thoroughly discussed (Fytas *et al.*, 2016; Hartmann and Young, 2001).

4.7.7 $d = 3$ Random-field neutron scattering line shapes

The great challenge faced when interpreting neutron scattering data for the random-field Ising critical behavior is the lack of a theoretical foundation

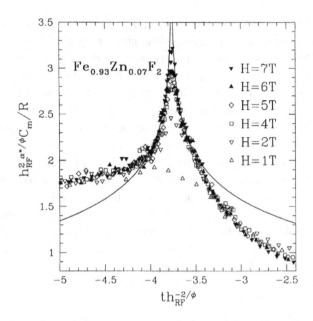

Fig. 4.77: The scaling of the magnetic specific heat of $Fe_{0.93}Zn_{0.07}F_2$ (Slanič and Belanger, 1998). The data at the smallest fields are rounded by the concentration gradient. The solid curves represent a symmetric, logarithmic divergence.

for the scattering line shapes that can be used to fit the data and extract estimates for the universal parameters such as the critical exponents ν, γ, and η and amplitude ratios associated with the fluctuation correlation length and susceptibility. The scattering intensity as a function of q is given by Belanger and Young (1991)

$$S(q) = \chi_S(q) + \chi_S^{\text{dis}}(q), \tag{4.29}$$

after correcting for instrumental resolution. Because the crystals have high crystalline quality and the scattering intensity at the $d = 3$ Bragg points is so large, extinction effects distort the temperature dependence of the Bragg scattering intensity. Neutron scattering, therefore, can only address scattering data outside the instrumental resolution ellipsoid, so the part of the disconnected susceptibility that represents the order parameter will not be part of the analysis. The behavior of the excluded data was addressed by the X-ray scattering covered in Section 4.4. Neutron scattering is ideal, however, for the range of q larger than the range accessible to the X-ray scattering. The two techniques are complementary in that regard.

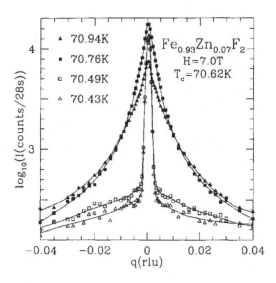

Fig. 4.78: Scattering line shapes for $Fe_{0.93}Zn_{0.07}F_2$ with $H = 7$ T very close to $T_C(H)$ (Slanič *et al.*, 1998). The line shapes below $T_C(H)$ are much smaller for $q \neq 0$, whereas the Bragg intensities are similar for temperatures just above and below $T_C(H)$.

Figure 4.78 shows scans for the $Fe_{0.93}Zn_{0.07}F_2$ crystal taken with $H = 7$ T. One striking feature of the data taken at this concentration, which is well above the vacancy percolation threshold concentration $x_v = 0.754$, is that the data are insensitive to the thermal protocol; FC and ZFC data are identical. As discussed above, hysteresis for $x > x_v$ is only observed for extremely small values of $|q|$ that are not included in the fitted neutron scattering data. A feature that stands out in Fig. 4.78 is the difference between the scattering intensities for $T > T_C(H)$ and $T < T_C(H)$. This is largely a result of a significant contribution to $\chi_S^{dis}(T)$ intensity transferring from short-range order ($|q| > 0$) above $T_C(H)$ to Bragg scattering ($q = 0$) below it. Even so, it will be shown below that the scattering line shape for $T < T_C(H)$ is far from Lorentzian-like.

The staggered susceptibility should have the form

$$\chi_S(q) = A^{\pm}\kappa^{\eta-2}f(q/\kappa), \tag{4.30}$$

and the disconnected staggered susceptibility for $|q| > 0$ should have the form

$$\chi_S^{dis}(q) = B^{\pm}\kappa^{\bar{\eta}-4}g(q/\kappa), \tag{4.31}$$

where the $+$ and $-$ superscripts refer to $T > T_C(H)$ and $T < T_C(H)$, respectively. The actual forms for $f(q/\kappa)$ and $g(q/\kappa)$ are not known from theory and can be different above and below $T_C(H)$, so the data are fit separately for $T > T_C(H)$ and $T < T_C(H)$. Assumptions must be made regarding the forms of $f(q/\kappa)$ and $g(q/\kappa)$ in order to fit the scattering data. One initial assumption is that $A^{\pm}$ and $B^{\pm}$ can be fit as if they are independent; we will return to that point later.

As always, a reasonable start is with the mean-field approximation, which for the random-field case, means $\chi_S(q)$ has a Lorentzian form

$$f(q) = \frac{1}{1 + q^2/\kappa^2}, \tag{4.32}$$

and $\chi^{\mathrm{dis}}(q)$ has a squared-Lorentzian form

$$g(q) = \frac{1}{(1 + q^2/\kappa^2)^2}, \tag{4.33}$$

for $q \neq 0$.

These forms work fairly well for $T > T_C(H)$, as discussed below, but they fail below the transition. To go beyond mean-field, in the same spirit as was done for the random-exchange behavior at $H = 0$, reasonable forms for $f(q/\kappa)$ can have the forms known for the pure Ising case, namely

$$f(q/\kappa) = \frac{(1 + \phi^2 q^2/\kappa^2)^{\eta/2}}{1 + \psi q^2/\kappa^2}, \tag{4.34}$$

for $T > T_C(H)$ and

$$f(q/\kappa) = \frac{(1 + \phi'^2 q^2/\kappa^2)^{\sigma + \eta/2}}{(1 + \psi' q^2/\kappa^2)(1 + \phi''^2 q^2/\kappa^2)^{\sigma}}, \tag{4.35}$$

for $T < T_C(H)$.

This still leaves the function $g(q/\kappa)$ to be defined for the fits, but adding a whole new set of parameters by introducing independent forms like those for $f(q/\kappa)$ would overwhelm any attempts at fitting the data, so additional approximations are made, namely that

$$g(q/\kappa) = f^2(q/\kappa), \tag{4.36}$$

and

$$\bar{\eta} = 2\eta, \tag{4.37}$$

resulting in the form for the scattering function

$$S(q) = A^{\pm}\kappa^{\eta-2}f(q/\kappa) + B^{\pm}\kappa^{2\eta-4}f^2(q/\kappa). \tag{4.38}$$

Setting $\bar{\eta} = 2\eta$ is a fairly good approximation (Fytas and Malakis, 2011; Fytas and Martín-Mayor, 2013; Fytas *et al.*, 2018; Schwartz and Soffer, 1986).

The random-exchange to random-field crossover happens as $|t| \to 0$ and as $|q| \to 0$. It is often safe to assume that, because the scattering line shape is dominated by the small $|q|$ and small $|t|$ intensities, the measured line shapes should approximate the asymptotic random-field line shape that dominates fits of the data. It is usual to ignore the crossover in $|q|$, as was the case, for example, in fits to the pure and random-exchange scattering line shapes. It is then possible that crossover in $|t|$ is observable as the fits to κ and χ yield power law behaviors in $|t|$ that change slope in logarithmic plots. In practice, it is hard to discern such crossover. What is more likely is that fits yield effective exponents that approach the universal values as the data become dominated by the asymptotic values when $|t|$ is sufficiently small or the randomness in the exchange is sufficiently large. Keeping this approximation in mind, there is still a problem in that there are simply too many parameters to fit if the usual procedure of fitting each scan and examining the parameters as a function of T is used. Instead, the data at all temperatures for $T > T_C(H)$ can be fit and, independently, all the scans at temperatures $T < T_C(H)$ can be fit simultaneously. This works using $\kappa = \kappa_0|t|^\nu$ and ignoring crossover effects, which are hopefully minor for small $|q|$ and $|t|$. Note that by setting $B^{\pm} = 0$, we recover the random-exchange form for $S(q)$ that was used in a similar way with excellent results for $H = 0$, as discussed in Section 4.6. Fitting all the temperatures simultaneously might hide random-exchange to random-field crossover effects, but it is not known how big a role this might have in the interpretation. The great advantage is that it does allow reasonable fits, as discussed below.

For the discussions of the data, it is important to note that, for $x > x_v$, the neutron scattering line shapes, unlike those from X-ray experiments, show no hysteresis. It does not matter whether the data are taken using ZFC, FC, or FH protocols. This contrasts the behavior for $x < x_v$ where severe hysteresis is observed in neutron scattering data, and analyses could only be done well above the equilibrium temperature $T_{\text{eq}}(H)$. Hence, for the large concentration samples, the line shapes can be considered to represent equilibrium conditions for the range of accessible $|q|$. The X-ray data described above do show hysteresis, but that hysteresis involves only length

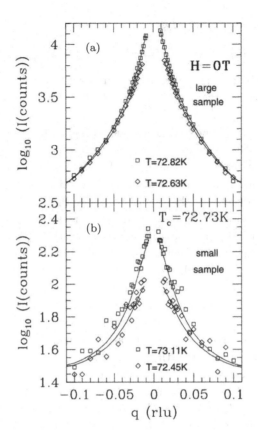

Fig. 4.79: Fe$_{0.93}$Zn$_{0.07}$F$_2$ line shapes for $H = 0$ (Slanič *et al.*, 2001).

scales not accessible because of the more limiting resolution in the neutron scattering experiments.

The scaling behavior of the line shapes in

$$S(q) = C(\kappa^{\eta-2}f^{\pm}(q/\kappa) + \text{ratio}^{\pm}(\kappa^{\eta-2}f^{\pm}(q/\kappa))^2) + B + Dq, \qquad (4.39)$$

where f^+ is the function used for $T > T_C(H)$ in the analysis of the $H = 0$ data with fitted line shape parameters, and f^- is the function used for $H = 0$ for $T < T_C(H)$. The fits are shown for Fe$_{0.93}$Zn$_{0.07}$F$_2$ in Figs. 4.79 and 4.80 for $H = 0$ using only ratio$^{\pm} = 0$ and in Figs. 4.81 and 4.82 for $H = 7$ T. The fitted parameters are in Tables 4.12 and 4.13.

Similar fits for $H > 0$ were made to neutron scattering data collected for $x = 0.87$ (Ye, 2003). Table 4.14 shows how the exponents η and ν vary

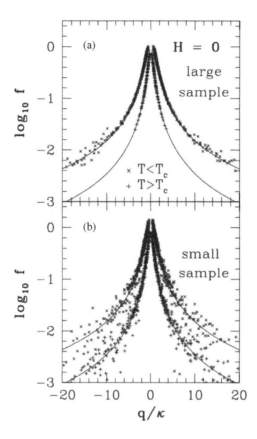

Fig. 4.80: The scaling behavior of the $Fe_{0.93}Zn_{0.07}F_2$ line shapes for $H = 0$, as described in the text (Slanič *et al.*, 2001).

with the applied field. It suggests that ν and η might be effective exponents that tend to increase with the strength of the random-field.

Experimentalists face an extraordinary challenge in characterizing the $d = 3$ random-field scattering data, including a lack of theoretical indications regarding parameters appropriate to the line shapes in Eqs. (4.32) and (4.33) that would be helpful in fits to the data. It is also not clear how to incorporate random-exchange to random-field crossover into the fitting of the line shapes to the data. What is clear from Table 4.14 is that the fitted value of ν increases with the random-field strength, which increases with the strength of the applied uniform field or with the increase in magnetic dilution.

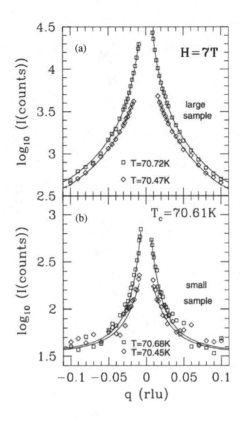

Fig. 4.81: $Fe_{0.93}Zn_{0.07}F_2$ line shapes for $H = 7$ T (Slanič *et al.*, 2001).

One question that remains is how the contributions from $f^{\pm}(q/\kappa)$ and $(f^{\pm}(q/\kappa))^2$ are related. That point is possibly addressed in part by examining the scattering in a completely different way. Figure 4.83 shows how the envelope of the line shape as $|t| \to 0$ should behave based on an underlying fractal structure for $d = 2$ and $d = 3$. The $d = 2$ case is a calculation, whereas the $d = 3$ case shows the behavior of scattering data for $Fe_{0.85}Zn_{0.15}F_2$ in an applied field $H = 10$ T for $T > T_C(H)$. In the latter case, the line shape is compared to scattering expected from the fractal structure of the magnetic ordering (Seppälä and Alava, 2001; Seppälä *et al.*, 2002; Ye *et al.*, 2004). Fitting the data to Eq. (4.34), the behavior $S(q) \to |q|^{-D_f}$ is observed as $t \to 0$, with $D_f = 2.53$, which is the fractal dimension appropriate for $d = 3$. This is clear from the figure where, as the transition is approached from above, the line shapes asymptotically conform

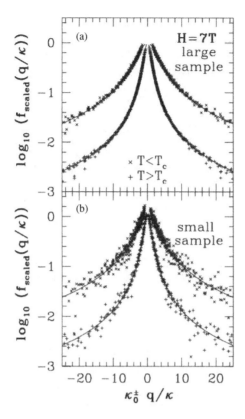

Fig. 4.82: The scaling behavior of the $Fe_{0.93}Zn_{0.07}F_2$ line shapes for $H = 7$ T, as described in the text (Slanič *et al.*, 2001).

to the $d = 3$ fractal shape. The curve shown for the experimental case has the instrumental resolution folded in. The value of $\nu = 1.20(5)$ obtained from the fits is similar to 2ν obtained using Eq. (4.39). This suggests that the fractal structure might be an intrinsic part of the critical behavior of the line shapes. The fits to a single line shape that incorporates the fractal structure suggest that the parameter ratio[+] in Eq. (4.39) is determined by the fractal nature of the magnetic structure.

A few recent exponent estimates from theory include: $\nu = 1.41(15)$ (Ahrens *et al.*, 2013); $\nu = 1.38(10)$ and $\eta = 0.5153(9)$ (Fytas and Martín-Mayor, 2013); $\gamma = 2.28(2)$ (Xiong and Xu, 2019); $\eta = 0.51(2)$, $\bar{\eta} = 1.01(3)$, $\gamma = 1.95(27)$, and $\bar{\gamma} = 3.92(54)$ (Fytas and Malakis, 2011); $\nu = 0.90(15)$ (Fernandez *et al.*, 2011); $\nu = 1.37(1)$ (Fytas *et al.*, 2016); $\nu = 1.38(2)$

Table 4.12: $d = 3$ Random-exchange and random-field Ising parameters (Slanič, 1998; Slanič *et al.*, 2001).

	Lorentzian	Modified	Modified	Modified
$|t|_{max}$	0.15	0.15	10^{-2}	3×10^{-3}
H	0	0	7 T	7 T
ν	0.70(3)	0.70(2)	0.88(5)	0.87(7)
κ^+	0.57(2)	0.56(2)	1.13(4)	0.95(17)
κ^+/κ^-	0.471(42)	0.496(28)	0.349(17)	0.342(87)
γ	1.34(6)	1.34(4)	1.58(13)	1.60(16)
χ^+/χ^-	4.8(9)	4.6(5)	3.8(2)	4.9(10)
η	0.079(10)	0.079(12)	0.20(5)	0.16(6)

Notes: The exponent $\gamma = (2 - \eta)\nu$ is calculated. The "Modified" fits are from the equations developed in this section.

Table 4.13: $d = 3$ Random-field line shape fitted parameters for $x = 0.93$ (Slanič *et al.*, 2001).

H	0	7 T	7 T
	$|t| < 0.15$	$|t| < 10^{-2}$	$|t| < 3 \times 10^{-3}$
σ	0.16(20)	0.67(50)	0.86(60)
ϕ	0.18(2)	0.16(4)	0.08(1)
ϕ'	0.18(2)	0.39(25)	0.36(30)
ϕ''	0.08(10)	0.31(25)	0.26(20)

(Theodorakis *et al.*, 2013); and $\nu = 1.32(7)$ and $\nu = 0.50(3)$ (Hartmann and Young, 2001). The different estimates for ν are probably a result of emphasizing the two contributions to the line shapes versus the most narrow part of the line shapes. A better guide to an accurate line shape to do the experimental analysis could be helpful in making a more meaningful comparison of the data fitted parameters to the results from simulations (Fytas and Martín-Mayor, 2013).

We can return to the early neutron scattering results (Belanger *et al.*, 1985b) for $x = 0.6 < x_v$, which were obtained using a modified Lorentzian plus squared-Lorentzian term. Although the data could only be fit for $T > T_N$, exponent values obtained, $\nu = 1.0 \pm 0.15$, $\gamma = 1.75 \pm 0.2$, and $\eta \approx 1/4$, are consistent with the more accurate results obtained for $x = 0.93$ because the error estimates are large. This suggests the critical exponent

Table 4.14: $d = 3$ Random-field Ising system $Fe_x Zn_{1-x} F_2$ with $x = 0.87$ and $x = 0.85$ fit above $T_C(H)$.

Ye (2020) $x = 0.87$				
	$H = 5.0\,T$	$6.9\,T$	$H = 8.5\,T$	$H = 10\,T$
$t_{min} - t_{max}$	0.01–0.3	0.008–0.4	0.01–0.4	0.01–0.35
$|q|_{min} - |q|_{max}$	0.006–0.3	0.004–0.3	0.006–0.3	0.006–0.25
η	0.44(26)	0.41(1)	0.43(1)	0.64(1)
ν	0.82(3)	0.94(1)	0.94(1)	1.08(1)
Ye *et al.* (2004) $x = 0.85$				
$t_{min} - t_{max}$				0.004–0.3
η				0.58(5)
ν				1.18(5)
Ye *et al.* (2004) $x = 0.85$ (fractal)				
$t_{min} - t_{max}$				0.004–0.06
ν				1.20(5)

Notes: As the concentration decreases, the strength of the random field increases. As the random field strengthens, the exponent ν increases. The last set of entries uses only a single term in the line shape and the result is a fit consistent with a fractal line shape, as shown in Fig. 4.83 while the exponent ν is nearly equal to the fit using the Lorentzian plus squared-Lorentzian line shape.

results, obtained above $T_{eq}(C)$, where metastable domains are not an issue, are not greatly affected by the vacancy percolation. The difficulty in reaching equilibrium on the longest length scales and the resulting metastable domains that form upon FC are consistent with the predicted activated dynamics of the random-field Ising model, as discussed in Section 4.7.3.

4.8 $d = 3$ XY Specific Heat Critical Behavior

High resolution critical behavior studies of the $d = 3$ XY universality class are not as common as they are for the Ising or Heisenberg universality classes. Two systems that have been studies using specific heat, thermal diffusivity and thermal conductivity are $CsMnF_3$ (Oleaga *et al.*, 2014) and $SmMnO_3$ (Oleaga *et al.*, 2012). Figure 4.84 shows the specific heat behavior measured in $CsMnF_3$. The critical exponent α and the amplitude ratio A^+/A^- values are listed in Table 4.15 along with theoretical values (Pelissetto and Vicari, 2002) and show excellent agreement.

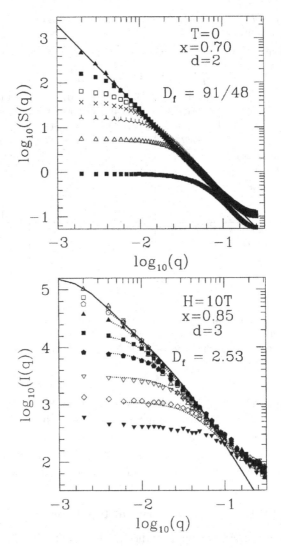

Fig. 4.83: Calculated fractal line shapes for $d = 2$ (upper figure) and measured line shapes fit to $d = 3$ scattering data (lower figure) (Ye *et al.*, 2004).

4.9 Experiments on Isotropic Magnets

4.9.1 *Optical birefringence in isotropic antiferromagnets*

The isotropic antiferromagnets $KNiF_3$ and $RbMnF_3$ were studied (Nordblad *et al.*, 1983a) using optical birefringence. For $T > T_N$, there is

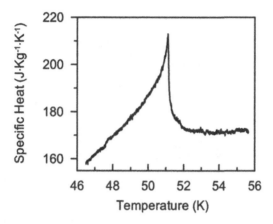

Fig. 4.84: The specific heat of the $d = 3$ XY system CsMnF$_3$ (Oleaga *et al.*, 2014). Reproduced with permission from IOP Publishing.

Table 4.15: The specific heat critical behavior of CsMnF$_3$ (Oleaga *et al.*, 2014) and SmMnO$_3$ (Oleaga *et al.*, 2012) compared to theoretical estimates from the list of results tabulated in (Pelissetto and Vicari, 2002).

CsMnF$_3$		
$\|t\|_{min}$	8.8×10^{-4} $(T < T_N)$	5.9×10^{-5} $(T > T_N)$
$\|t\|_{max}$	9×10^{-2} $(T < T_N)$	8.7×10^{-2} $(T > T_N)$
α	$-0.016(4)$	
A^+/A^-	1.09	
SmMnO$_3$		
$\|t\|_{min}$	1.9×10^{-3} $(T < T_N)$	2.4×10^{-4} $(T > T_N)$
$\|t\|_{min}$	8.3×10^{-2} $(T < T_N)$	8.6×10^{-2} $(T > T_N)$
α	$-0.017(1)$	
A^+/A^-	1.07	
Theory		
α	-0.01 to -0.02	
A^+/A^-	1.05	

no signal because the system is essentially isotropic. Even when the magnetic system orders and breaks the local symmetry, the crystal overall shows no birefringence without some external influence to define the optical axes. Figure 4.85 shows two ways this was done. Data taken with uniaxial pressure applied to KNiF$_3$ are shown in the upper panel. The strength of the signal increases, but saturates at a pressure $P = 0.1$ kbar. In the lower panel, a magnetic field was applied. The logarithm of Δn for a pressure

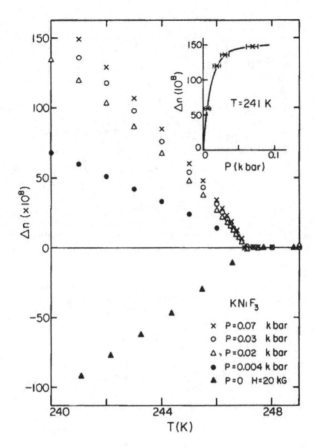

Fig. 4.85: The optical birefringence of KNiF$_3$ with applied uniaxial pressure and with an applied field (Nordblad *et al.*, 1983a).

$P = 0.03$ kbar is plotted as a function of the logarithm of the reduced temperature $|t|$ in Fig. 4.86. A fit to the data yields a value $\tilde{\beta} = 0.83(1)$. From Section 3.3.4, we use the expression $\tilde{\beta} = 2 - \phi - \alpha$, where α is the $d = 3$ Heisenberg specific heat exponent and ϕ is the crossover exponent. Using the values $\alpha = -0.14(1)$ from the table in the next section and $\phi = 1.25$ gives a value $\tilde{\beta} = 0.89(1)$, which is a bit higher than the experimental value. The value $\alpha = 0.11(2)$ (Marinelli *et al.*, 1996) yields a smaller exponent $\tilde{\beta} = 0.86(2)$. Very similar experimental results were obtained (Ferre *et al.*, 1982) for the Heisenberg systems RbMnF$_3$, KNiF$_3$, and KCoF$_3$, with exponents $\tilde{\beta} = 0.84(4)$, $0.82(2)$ and $0.82(2)$, respectively.

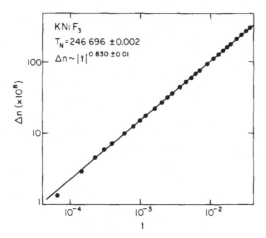

Fig. 4.86: The behavior of KNiF$_3$ with an applied pressure $P = 0.1$ kbar (Nordblad *et al.*, 1983a).

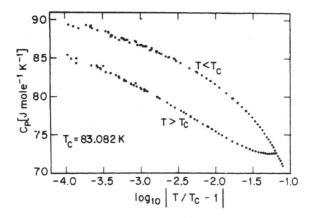

Fig. 4.87: Specific heat critical behavior of the isotropic antiferromagnet RbMnF$_3$ (Kornblit and Ahlers, 1973).

4.9.2 The specific heat critical behavior of isotropic magnets

The specific heat of RbMnF$_3$ was measured (Kornblit and Ahlers, 1973), with data approaching $|t| = 10^{-4}$, as shown in Fig. 4.87. The negative curvature, corresponding to $\alpha < 0$, is apparent across a wide range of $|T|$.

The specific heat critical behavior of the isotropic antiferromagnet KNiF$_3$ was measured (King *et al.*, 1984) using the capacitance technique

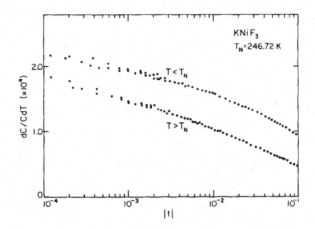

Fig. 4.88: Specific heat critical behavior of the isotropic antiferromagnet KNiF$_3$ measured using the capacitance technique (King *et al.*, 1984).

Table 4.16: The specific heat critical behavior of KNiF$_3$ (King *et al.*, 1984) RbMnF$_3$ (Kornblit and Ahlers, 1973).

	KNiF$_3$ (dC/dT) (King *et al.*, 1984)	RbMnF$_3$ (Kornblit and Ahlers, 1973)
	$10^{-4} < |t| < 0.02$	$3 \times 10^{-4} < |t| < 0.06$
α	$-0.151(4)$	$-0.137(4)$
A^+/A^-	$1.56(3)$	$1.40(4)$

Source: Reproduced with permission from American Physical Society.

described in Section 3.4. The results are comparable to the specific heat measurements shown in Fig. 4.87, but with a significantly smaller background. A fit to the critical behavior yields the results in Table 4.16 from the specific heat and capacitance techniques. The critical behavior has also been determined from specific heat and thermal diffusivity (Marinelli *et al.*, 1996) with the results $\alpha = -0.11(1)$ and $\alpha = -0.11(2)$, respectively, and $A^+/A^- = 1.27(9)$ for both, but the data were not taken as close to the transition as the experiments listed in the table. Overall, the results are in good agreement with the theoretical estimates listed in Tables 2.1 and 2.2.

Chapter 5

Domains, Excitations, and Spin-Glass-Like Behaviors

5.1 Domain Structure Dynamics at Low Temperature

In this section, some aspects of domain formation and domain dynamics will be further explored, particularly at low temperatures. Domains have played a central role in the discussions in previous sections regarding the random-field Ising model as realized in dilute antiferromagnets. Although, upon ZFC, a phase transition exhibiting critical behavior belonging to a new universality class has been realized for $d = 3$ measurements in an applied field, metastable FC domain structures are present even a low temperatures for concentrations below the vacancy percolation threshold x_v. An early work (Villain, 1984) provided arguments for why the FC thermal cycling cannot produce the long-range ordered system that is achieved under the ZFC procedure, based on the difficulty of the system in overcoming the pinning effect of the random-field and the fractal-like structure of the domains. The argument, formulated for a pure Ising magnet with random fields applied, introduces a typical radius that varies as $\ln(t/\tau)$, where t is time and τ varies with the correlation length as $\ln \tau \propto \xi^\theta$ (Fisher, 1986; Villain, 1984). Although metastable domains are not observed for $x > x_v$ at low temperatures, hysteresis is observed close to the transition temperature. Fractal percolation structures influence the critical behavior, as discussed in the previous chapter. Ground state calculations provide insight into the formation of domain structures in the random-field Ising models in $d = 2$ (Nowak *et al.*, 1996; Seppälä and Alava, 2001; Seppälä *et al.*, 1998) and $d = 3$ (Glaser *et al.*, 2005; Nowak and Usadel, 1992a; Seppälä *et al.*, 2004, 2002) dilute antiferromagnets.

For $d = 2$, the low temperature state with random fields is a domain state. After turning off the field, the domain wall dynamics have been studied in $Rb_2Co_{0.85}Mg_{0.15}F_4$ (Kleemann et al., 1998). For low temperatures, the behavior is described by dynamically coupled domains relaxing exponentially with

$$\delta M(t) \propto \exp(-t/\tau_\infty),\qquad(5.1)$$

with $t_\infty \approx 10^{-12}$s (Chamberlin, 1994). For higher temperatures, the generalized power law (Staats et al., 1998)

$$\delta M(t) = \delta M_0 \exp(-b(T\ln(t/\tau))^y)\qquad(5.2)$$

works well.

For $d = 3$, FC metastable domains are frozen in at low T for $x < x_v$. This provides an interesting mechanism for producing domain wall structures at low temperatures that can be studied by turning off the field and measuring the remanent magnetic moment as a function of time. A number of studies have examined the evolution of the domain walls once the field is removed. The excess magnetization ΔM at the domain walls is characterized by

$$\Delta M \propto H^m,\qquad(5.3)$$

where H is the applied field and m is an exponent. At very low temperature, $m = \nu_H$, a correlation length exponent that can be estimated from neutron scattering line shapes assuming mean-field Lorentzian plus squared-Lorentzian forms, with resulting values 2.2(1) and 2.1(1) (Belanger et al., 1985a). The domain size of typical domains is then given by

$$R \approx H^{-\nu_H},\qquad(5.4)$$

and early modeling assumed that $m = \nu_H$ for smooth domain walls. A more recent analysis (Mattsson et al., 2000) at higher temperatures included the fractal nature of the domain interfaces and proposed that

$$m = \nu_H + (2D_f + 2 - 2d)/(d - 2),\qquad(5.5)$$

where D_f is a fractal dimension.

When the field is removed, the excess magnetization ΔM decays as a remanent magnetization μ. A number of time dependencies have been proposed and used to analyze the time dependence of μ, such as (Djurberg *et al.*, 1994; Leitão *et al.*, 1988a,b; Nattermann and Vilfan, 1988)

$$\mu = AH^x (T \ln(t/t_0))^{-\Psi}, \qquad (5.6)$$

or (Han and Belanger, 1992; Han *et al.*, 1992)

$$\mu \propto \exp(-b(\ln(t/\tau)^\beta)), \qquad (5.7)$$

or (Lederman *et al.*, 1992, 1993)

$$\mu \propto (\ln(t/\tau))^{-1}. \qquad (5.8)$$

A form of Eq. (5.7) was proposed by (Mattsson *et al.*, 2000),

$$\mu \propto H^m (1 - c(T \ln(t/\tau_0)^a)), \qquad (5.9)$$

where τ_0 is a microscopic relaxation time of order 10^{-12} s. This form removes much of the T and H dependence of the parameters in the behavior seen using other forms. Measurements were made using $Fe_{0.6}Zn_{0.4}F_2$ and are shown in Fig. 5.1. The value $m = 2.9(1)$ was determined, in agreement with the value $m = 3.2(3)$ found previously (Lederman *et al.*, 1992) for $x = 0.46$.

Using the values $m = 2.9$ and $\nu_H = 2.1$, the fractal dimension $D_f = 2.4$ was determined for $H > 2.5\,T$. The FC fractal domain structure is largely determined upon cooling near the phase transition and the scattering line shapes just above the transition exhibits possible fractal structures that were fit using the fractal dimension $D_f = 2.53$ (see Section 4.7.7). The experimental value $D_f = 2.4$ for the relaxation measurements is compatible with the scattering value $D_f = 2.53$, considering the uncertainties in the parameters used to determine the values.

For small fields $H < 2.5$ T, the behavior of μ versus T in $Fe_{0.47}Zn_{0.53}F_2$ is dominated (Kushauer *et al.*, 1994) by a moment that is largely independent of H, as seen in Fig. 5.2. It is believed that the source is piezomagnetism in strongly anisotropic systems or a magneto-elastic coupling (Lederman *et al.*, 1990) that can be attributed to the rutile structure of pure and diluted FeF_2.

A small remanent magnetization was also observed in the $Mn_xZn_{1-x}F_2$ low-anisotropy system at extremely small fields, as shown in Fig. 5.3.

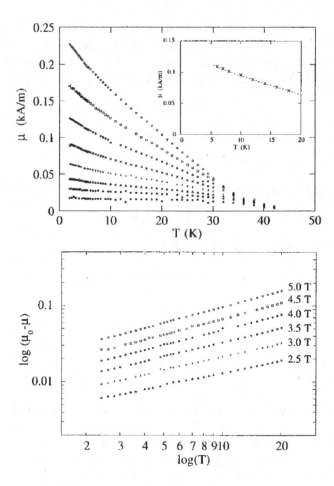

Fig. 5.1: The remanent magnetization μ versus T (upper panel) after cooling to low temperatures in fields, from the top, of $H = 5$, 4.5, 4, 3.5, 3, 2.5, 2, and 1 T and then removing the field, and the logarithm of $\mu(0) - \mu$ versus the logarithm of T (lower panel) for $2.5 \leq H \leq 5$ T, measured in $Fe_{0.6}Zn_{0.4}F_2$ (Mattsson *et al.*, 2000). The insert shows a comparison of the data in the main figure along with μ versus T after cooling to the temperatures shown and waiting 30s for $H = 4$ T. Reproduced with permission from American Physical Society.

Although piezomagnetism is a possible source of the remanent magnetization, its universal behavior is shared with other low-anisotropy systems where the crystal lattice does not allow that mechanism (Becerra *et al.*, 1995; Palacio *et al.*, 1994). A review of impurity-induced effects in low anisotropy antiferromagnets is given in Palacio *et al.* (1994).

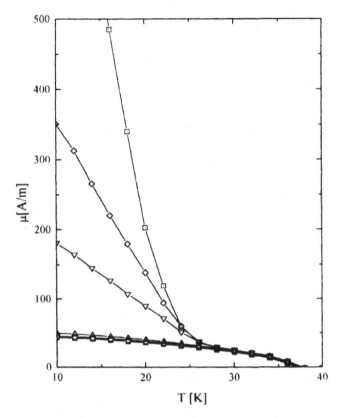

Fig. 5.2: The remanent magnetization μ versus T upon heating in zero field after FC in fields, from the upper to the lower curves, of $H = 4$, 3.5, 3, 2, 1.5, 0.03, 0.01, and 0.001 T (Kushauer *et al.*, 1994). The lower fields give results that are essentially independent of H. Used with permission, American Physical Society.

5.2 The Phase Diagram of Anisotropic Antiferromagnets Above the Magnetic Percolation Threshold Concentration

Both the strongly anisotropic $Fe_xZn_{1-x}F_2$ and weakly anisotropic $Mn_xZn_{1-x}F_2$ systems have a magnetic percolation threshold concentration $x_p = 0.246$ for the dominant J_2 interaction. The other small interactions J_1 and J_2 could slightly complicate the percolation behavior very close to x_p. The transition temperature in zero applied field goes to zero near the percolation threshold concentration. The magnetic specific heat, measured using pulsed heat (Sousa *et al.*, 2010) and birefringence (Belanger, 1981)

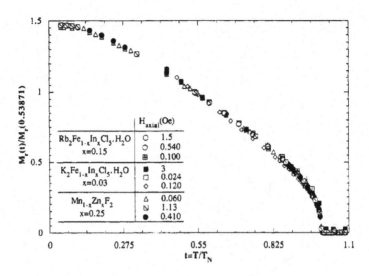

Fig. 5.3: The universal behavior of the remanent magnetization observed in low-anisotropy systems, including $Mn_{0.75}Zn_{0.25}F_2$ (Palacio *et al.*, 1994). Reproduced with permission from IOP Publishing.

techniques, are shown as a function of x in Fig. 5.4. Both techniques yield the same behavior; the magnetic specific heat decreases rapidly with magnetic dilution and the critical part becomes very small relative to the noncritical contributions. For the birefringence case, $d(\Delta n)/dT$ versus T for ZnF_2 is also shown to emphasize how small the lattice contribution is.

At x_p, there is no phase transition, but there is a broad magnetic peak centered around $T = 20$ K from the short-range magnetic correlations. The nature of the correlations were investigated (Belanger and Yoshizawa, 1993) using neutron scattering for $x = x_p = 0.25$ and $x = 0.27$. No resolution-limited Bragg scattering peak was observed in either case, showing that neither achieved long-range order. The data could not be fit to a simple Lorentzian line shape, but a Lorentzian plus squared-Lorentzian line shape used in the random-field experiments worked well. The squared-Lorentzian amplitude was significantly larger for $x = 0.27$, as shown in the top panel of Fig. 5.5, which might reflect that the nearly ordered state below $T = 10$ K has considerable domain structure. Both crystals show the inverse correlation length κ decreasing with T until $T = 10$ K, as shown in the bottom panel of Fig. 5.5. The minimum reached for $x = 0.27$ is significantly smaller than that achieved for $T = 0.25$, indicating that the $x = 0.27$ crystal has more antiferromagnetic order at low T. Small frustrations in the

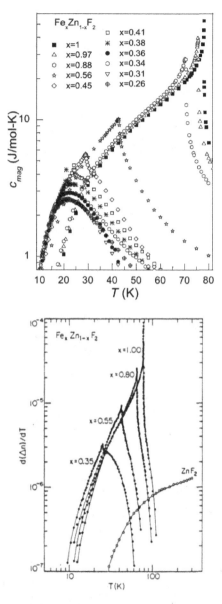

Fig. 5.4: The specific heat (Sousa *et al.*, 2010) and $d(\Delta n)/dT$ (Belanger, 1981) versus T of $Fe_x Zn_{1-x}F_2$. Both the overall magnitude and the phase transition contribution rapidly diminish as x decreases; the latter disappears at x_p. The significant phonon background contribution to the specific heat was removed by subtracting the specific heat of ZnF_2. The $d(\Delta n)/dT$ behavior was extended to the percolation threshold in Fig. 3.5. Reproduced figure used with permission from IOP Publishing.

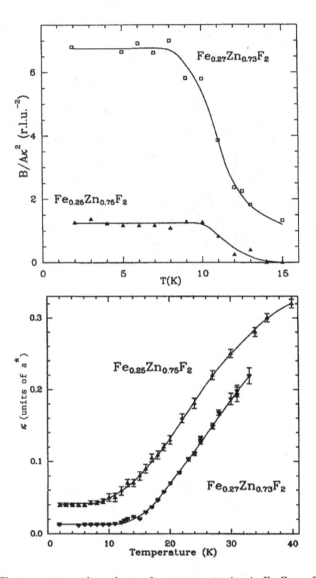

Fig. 5.5: The temperature dependence of neutron scattering in $Fe_x Zn_{1-x} F_2$ near the magnetic percolation threshold (Belanger and Yoshizawa, 1993). The data for $x = 0.27$ are fit with a significant squared-Lorentzian term, although these are zero-field measurements. The fits at $x = 0.25$ showed a smaller squared-Lorentzian contribution. The inverse correlation length, κ, is smaller at low temperatures in the $x = 0.27$ crystal.

interactions could prevent the fractal structure of the $x = 0.25$ magnetic lattice from approaching long-range order at $T = 0$.

A series of experiments (da Cunha *et al.*, 1990; de Araújo *et al.*, 1991; Montenegro *et al.*, 1988a,b; Montenegro *et al.*, 1989; Rezende *et al.*, 1988) on $Fe_{0.25}Zn_{0.75}F_2$ established the phase diagram at $x = x_p$. Simulations confirmed the behavior (Barbosa *et al.*, 2000, 2003; de Lima *et al.*, 2012). Magnetometry measurements at $x = 0.25 = x_p$ reveal hysteresis and an equilibrium boundary that scales as $T_i(0) - T_i(H) \propto H^{2/\phi}$, where $\phi = 3.4$, an exponent that is similar to that of the de Almeida–Thouless boundary for spin-glasses (Binder and Young, 1986; de Almeida and Thouless, 1978), as shown in Fig. 5.6. A canonical spin-glass shows a characteristic frequency dependence where the peak in the susceptibility occurs at a temperature that becomes smaller as the frequency decreases. For the canonical spin-glass, the peak temperature converges to a finite temperature as the frequency decreases. However, as seen in Fig. 5.7, no such temperature was found for $Fe_{0.25}Zn_{0.75}F_2$ (Jonason *et al.*, 1997), suggesting that, although many features are found to be common for the canonical spin-glass and percolation threshold Ising systems, they are in fact distinct with respect to having a finite T transition at low frequencies. A similar result was found for $Fe_{0.3}Mg_{0.7}Cl_2$ (Bertrand *et al.*, 1988).

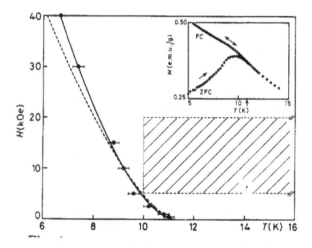

Fig. 5.6: The equilibrium boundary for $Fe_{0.25}Zn_{0.75}F_2$. The inset shows the ZFC-FC hysteresis for $H = 500$ Oe. The dashed line represents $\phi = 3$ and the solid line, which fits the data well, is $\phi = 3.4$. Note the exponent value is far from the random-field exponent $\phi_{rf} = 1.42$ (Montenegro *et al.*, 1989). Reproduced with permission from IOP Publishing.

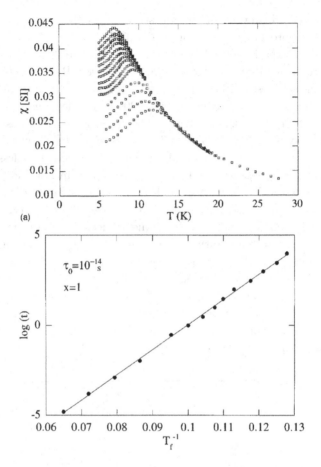

Fig. 5.7: The temperature dependence of the ac-susceptibility of $Fe_{0.25}Zn_{0.75}F_2$ in the upper panel for frequencies from bottom to top of 1.6×10^{-5}, 1.6×10^{-4}, 1.3×10^{-3}, 0.01, 0.3, 1, 3, 10, 30, 100, 300, 10^3, 3×10^3, and 10^4 s and the logarithm of t versus $1/T_f$ in the lower panel (Jonason *et al.*, 1997).

For concentrations sufficiently above x_p, the behavior of $Fe_xZn_{1-x}F_2$ and the less anisotropic $Mn_xZn_{1-x}F_2$ in a small field is that of the random-field Ising model, with a phase boundary that scales with $H^{2/\phi_{rf}}$, where $\phi_{rf} \approx 1.42$ is the random-field crossover exponent (see Section 4.3.2). Two effects lead to an instability of the random-field transition. First, the size of the random field increases with H and with dilution. Second, the magnetic lattice becomes fragile as it is diluted because of the percolating vacancy lattice. Both of these contribute to the breakdown of long-range order with increasing H. It has been shown (Belanger *et al.*, 1991; Bensamka *et al.*, 1987; de Araújo *et al.*, 1991; Montenegro *et al.*, 2000a,b, 1992, 1994, 1991, 1998, 1999, 2000c, 1988b, 1995; Rosales-Rivera *et al.*, 2000; Wong *et al.*, 1985) that above the region where the system orders, the boundary changes curvature and scales as $H^{2/\phi}$, where $\phi = 3.4$ is the same exponent that describes the entire boundary for $x = x_p$. Simulations show similar effects (Barbosa *et al.*, 2005; Nowak and Usadel, 1991). Figure 5.8 shows this behavior for two concentrations of $Fe_x Zn_{1-x} F_2$ for x well above x_p.

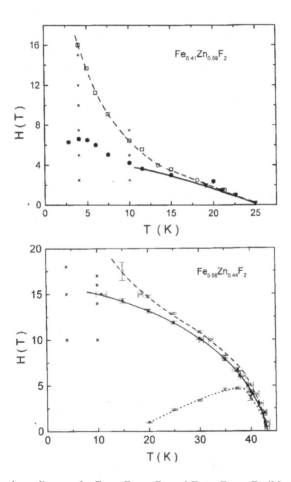

Fig. 5.8: The phase diagram for $Fe_{0.41} Zn_{0.59} F_2$ and $Fe_{0.56} Zn_{0.44} F_2$ (Montenegro *et al.*, 2000c). The behavior at lower fields is typical of the random-field Ising model but at the higher fields the phase boundary exhibits spin-glass-like behavior. Reproduced with permission from Brazilian Journal of Physics.

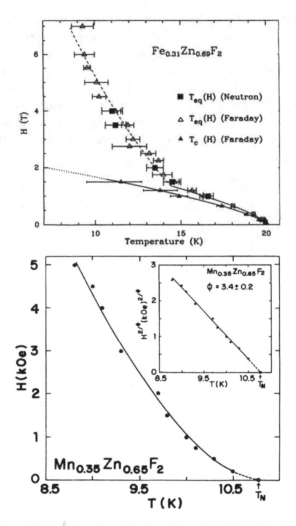

Fig. 5.9: The equilibrium boundary for $Fe_{0.31}Zn_{0.69}F_2$ (Belanger *et al.*, 1991) and $Mn_{0.35}Zn_{0.65}F_2$ (Montenegro *et al.*, 1995). The low field behavior for $Fe_{0.31}Zn_{0.69}F_2$ is the random-field behavior and the high field behavior is spin-glass-like. For $Mn_{0.35}Zn_{0.65}F_2$, the entire phase boundary is spin-glass-like. Reproduced with permission from American Physical Society.

Figure 5.9 shows the behavior just above x_p and for $Fe_xZn_{1-x}F_2$ with $x = 0.31$ and $Mn_xZn_{1-x}F_2$ with $x = 0.35$. Just above x_p, the highly anisotropic system shows random-field behavior at small H and spin-glass-like behavior at larger fields, but the weakly anisotropic system, with a slightly higher magnetic concentration, shows the spin-glass-like behavior for all fields. It is likely that the anisotropy in $Mn_xZn_{1-x}F_2$, which is dipolar in origin, is not sufficient to produce the typical random-field Ising behavior at low fields. This is partly the result of the dipolar fields at each Mn site that are generally not along the c axis as they are in the pure case. The point at which the behavior crosses from random-field to spin-glass-like is shown in Fig. 4.67 of Chapter 4 for $Fe_xZn_{1-x}F_2$. The features of the phase diagram are similar for all $x_p < x < x_v$.

5.3   Excitations Near the Magnetic Percolation Threshold Concentration

Although a thermal magnetic phase transition does not occur at the percolation threshold concentration, x_p, magnetic transitions can take place at larger magnetic concentrations and percolation effects are part of the descriptions. In this section, some aspects of the magnetic behavior at and just above to the percolation threshold concentration will be described. Spin wave excitations in pure antiferromagnets at low temperatures are magnons that are narrow in energy and can vary in energy as the wavelength changes; the dispersion can be used to determine the parameters of Hamiltonians used to model the systems. Upon dilution of the magnetic sites, isotropic antiferromagnets exhibit fracton spin waves that are fundamentally different from magnons because they must propagate on the fractal magnetic lattice near the percolation threshold concentration. The signature of fractons is that the excitations are narrow in energy at long wavelengths, but are broad for short wavelengths because they are sensitive to the variations in the local environments. Fracton effects are seen in isotropic systems very close to the percolation threshold. Systems that do not satisfy these requirements will not be covered in this section as representing fracton excitations. The magnetic lattice loses its fractal nature quickly as the concentration deviates from x_p. Anisotropic systems also show changes in the excitation spectra when the concentration closely approaches the percolation threshold concentration. In this case, the excitations are well-described as Ising cluster excitations that can be observed above the percolation threshold concentration. However, anomalous spin diffusion takes place close to x_p, as described below. The measurements of isotropic and anisotropic excitations near the percolation threshold will be described for $d = 2$ and $d = 3$ systems. For these, the percolation threshold concentrations relevant to the dominant interaction in the systems discussed below are $x_p = 0.593$ for the $d = 2$ square lattice, $x_p = 0.312$ for the simple cubic lattice, and $x_p = 0.246$ for the body-centered tetragonal lattice (same as body-centered cubic lattice) (Stauffer and Aharony, 1994).

5.3.1   *Fracton excitations in isotropic diluted $d = 2$ and $d = 3$ antiferromagnets near the magnetic percolation threshold concentration*

The most studied examples of excitations in isotropic diluted $d = 2$ and $d = 3$ antiferromagnets near the magnetic percolation threshold concentration have been done using the $d = 3$ antiferromagnet $RbMn_xMg_{1-x}F_3$ (Ikeda *et al.*, 1994, 1995b, 1998a,b; Itoh *et al.*, 2007, 2011, 2009) and the $d = 2$ (Itoh *et al.*, 1998, 2011, 2012) antiferromagnet $Rb_2Mn_xMg_{1-x}F_4$, with concentrations close to the appropriate threshold concentration x_p of the magnetic Mn component, as listed above.

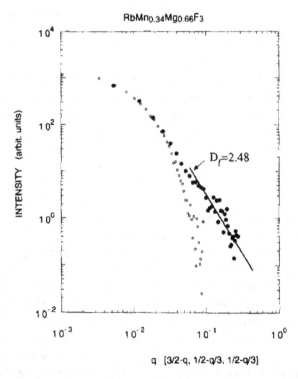

Fig. 5.10: The elastic neutron scattering intensities for $RbMn_{0.34}Mg_{0.66}F_3$ (solid) and the instrumental resolution (open) at low temperature, showing the fractal structure (Ikeda *et al.*, 1995b). Reproduced with permission from Elsevier.

Fracton excitations are expected to differ from spin waves that are observed in the pure systems (see Chen and Landau, 1993; Nakayama *et al.*, 1994 and references therein) and should be observed for wave vectors $q > q_G = 1/\xi_G$, where $\xi_G = a(x - x_p)^{-v_G}$, where a is a lattice constant and v_G is an exponent that is slightly less than 0.9 for $d = 3$ and 4/3 for $d = 2$ (Stauffer and Aharony, 1994). The elastic magnetic scattering for the spin-wave $q < q_G$ region is resolution limited but reflects the fractal lattice with

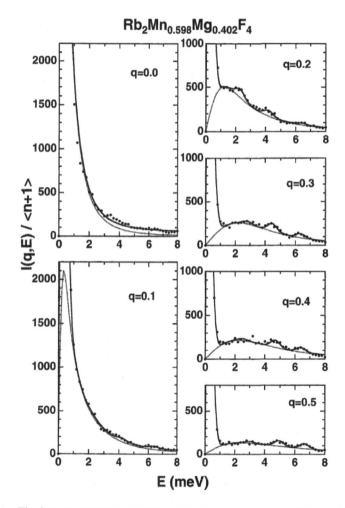

Fig. 5.11: The low temperature neutron scattering excitation spectrum for the $d = 2$ antiferromagnet $Rb_2Mn_{0.598}Mg_{0.402}F_4$ showing cluster excitation peaks and the broad fractal feature (Itoh *et al.*, 1998). Reproduced with permission from the Physical Society of Japan.

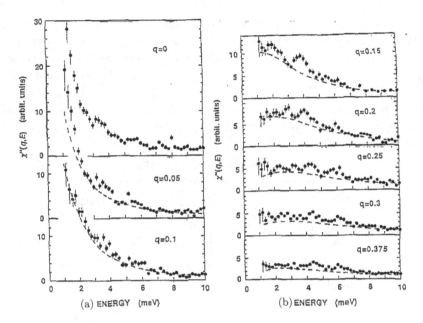

Fig. 5.12: The low temperature neutron scattering excitation spectrum for the $d = 3$ antiferromagnet RbMn$_{0.39}$Mg$_{0.61}$F$_3$ showing similar features as the $d = 2$ case in Fig. 5.11 (Ikeda *et al.*, 1994). Reproduced with permission from IOP Publishing.

$q \propto q^{-D_f}$ for $q > q_G$, where D_f is the fractal dimension equal to $91/48$ for $d = 2$ and approximately 2.48 for $d = 3$. For example, Fig. 5.10 shows the elastic scattering for the $d = 3$ system RbMn$_{0.34}$Mg$_{0.66}$F$_3$, which is close to x_p (Ikeda *et al.*, 1995b). The dynamical structure factor $S(q, \omega)$ has been calculated (Terao and Nakayama, 1995; Yakubo *et al.*, 1994), and many features have been determined in high resolution neutron scattering measurements for $d = 3$ in RbMn$_{0.598}$Mg$_{0.402}$F$_3$ and $d = 2$ in Rb$_2$Mn$_{0.4}$Mg$_{0.6}$F$_4$ and conform well to the theoretical expectations. Figures 5.11 (Itoh *et al.*, 1998) and 5.12 (Ikeda *et al.*, 1994) show the inelastic neutron scattering for $d = 2$ and $d = 3$, respectively. The general features, other than the concentrations are similar.

5.3.2 *Excitations in anisotropic $d = 2$ and $d = 3$ diluted antiferromagnets near the magnetic percolation threshold concentration*

Whereas pure FeF$_2$ exhibits spin wave dispersions, excitations in magnetically diluted Fe$_x$Zn$_{1-x}$F$_2$ well above x_p appears to show local spin cluster

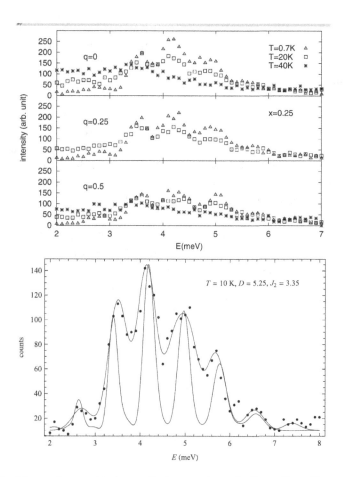

Fig. 5.13: The neutron scattering excitation spectrum for $Fe_{0.25}Zn_{0.75}F_2$ at the zone center $q = 0$, at $q = 0.25$, and at the zone boundary $q = 0.5$ (upper panel) and the calculated energy spectrum for $q = 0.5$ at $x = 0.31$ compared to the data and fit to the data at $T = 10$ K (lower panel) (Álvarez *et al.*, 2012). The energy spectrum is well described by interactions of spins with neighboring spins numbering between 0 and 8 with appropriate probabilities. Magnons play a minor role, if any, in this anisotropic antiferromagnet.

excitations as well as spin waves (Paduani *et al.*, 1994; Rodriguez *et al.*, 2007; Satooka *et al.*, 2002). Neutron scattering measurements (Álvarez *et al.*, 2012) very close to the percolation threshold concentration $x_p = 0.246$, made using $Fe_{0.31}Zn_{0.69}F_2$ and $Fe_{0.25}Zn_{0.75}F_2$, are shown in Fig. 5.13. Both crystals show an energy spectrum with multiple peaks at all wavelengths. The latter crystal is particularly interesting because the spins are on a well defined fractal lattice. The excitations have been well modeled (Álvarez *et al.*, 2012) using interactions similar to those listed in Table 4.1 for pure FeF_2 and accounting for the probabilities of neighboring spins ranging in number from 0 to 8. The results for $x = 0.31$ and $T = 10$ K are shown in the lower panel of Fig. 5.13. The agreement between the experimental and calculated spectra shows that the excitations are well understood as localized spin excitations with little magnon character.

For $d = 2$, four peaks were observed (Ikeda, 1994) using the anisotropic crystal $Rb_2Co_{0.58}Mg_{0.42}F_4$, which has a concentration close to x_p. The highest energy peak corresponds to the energy peak in $RbCoF_4$ and the lower energy peaks correspond roughly to 3/4, 1/2, and 1/4 of the highest energy and persist to temperatures comparable to T_N of pure Rb_2CoF_4 (Ikeda *et al.*, 1995b). High-resolution inelastic neutron scattering measurements on a $d = 2$ anisotropic crystal $Rb_2Co_{0.6}Mg_{0.4}F_4$ crystal revealed single-spin anomalous diffusion on the fractal percolating magnetic lattice by measuring the self correlation function (Ikeda *et al.*, 1995a).

Chapter 6

Experiments on Pure Magnets with Frustration

Magnetic order in Ising and Heisenberg antiferromagnets was discussed in preceding chapters where the universality class governing the critical behavior depended only on the dimensionality of the lattice and spin degrees of freedom. Quenched randomness in the spin interactions introduced a new random-exchange universality class. With the application of random fields in Ising systems, the transition is either destroyed ($d = 2$) or, in the case of the $d = 3$ Ising spin system, the critical behavior is modified by the introduced frustration and represents the random-field universality class. In this chapter, antiferromagnets without quenched randomness, but with frustration intrinsic to the magnetic lattice geometry or interactions, will be shown to represent new universality classes with unique critical behaviors. The first part will be a discussion of the stacked triangular lattice antiferromagnet $CsMnBr_3$ which has XY spins that are frustrated by the triangular geometry of the lattice; it will be shown to belong to the chiral universality class. Two other distinctly different chiral systems, holmium and VF_2, will be discussed in the following section. Finally, returning to the stacked triangular lattice, the frustrated Ising spins of $CsCoBr_3$ that belong to another universality class will be discussed. These are all examples of the different kinds of behaviors that can be introduced by frustration in the absence of quenched disorder.

One simple way to introduce frustration is to have antiferromagnetic interactions on a lattice with three ordering sublattices in a triangular arrangement. Triangular antiferromagnets were reviewed by Collins and Petrenko (1997). Phase transitions in two such cases will be discussed here: spins with XY anisotropy in $CsMnBr_3$ and spins with Ising anisotropy in

CsCoBr$_3$. In either case, the three sublattices cannot simultaneously order antiparallel with respect to their neighboring sublattices, despite the antiferromagnetic interactions between each sublattice. The two examples will be discussed here because they have been extensively studied and show interesting phase transitions in experiments and simulations.

6.1   The XY Stacked Triangular Lattice

The stacked triangular lattice magnet CsMnBr$_3$ consists of strongly interacting XY spins along chains arranged so that each spin forms a triangular configuration with neighbors on adjacent chains. The $d = 1$ chains would not order on their own, but a weak antiferromagnetic interaction between chains causes them to order as a $d = 3$ system. However, the triangular planes perpendicular to the chains do not allow adjacent spins to order antiferromagnetically with neighboring spins in the plane. Figure 6.1 shows the possible compromise arrangements of three XY spins on adjacent chains. The spins order at an angle of 120° with respect to their neighbors because it is impossible for all three sublattices to simultaneously order antiparallel to each other. There are two distinct chiral (Kawamura, 1986; Villain, 1977) configurations shown; a simple rotation cannot transform one into the other. As the system is cooled below the transition temperature, the symmetry is broken as the system chooses one of the two distinct configurations. The same frustration occurs throughout the triangular plane. Efforts to characterize the critical behavior of the proposed chiral universality class by experimentalists and theorists have been reviewed extensively (Kawamura, 1998; Pelissetto *et al.*, 2001).

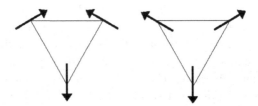

Fig. 6.1: The two chirality spin configurations of the stacked triangular lattice. The three antiferromagnetically interacting sublattices cannot order antiparallel (180°) to each other. Instead, the spins on the left rotate +120° moving counter-clockwise around the triangle and the spins on the right rotate +120° moving clockwise around the triangle; the former is referred to as having left-handed chirality and the latter as having right-handed chirality. One configuration cannot be transformed into the other by a simple rotation; they are two discrete states in the same sense that an Ising spin has two distinct states.

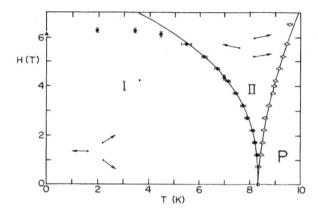

Fig. 6.2: The phase diagram of CsMnBr$_3$ showing the tetracritical point at $H = 0$ (Gaulin *et al.*, 1989). Reproduced with permission from American Physical Society.

Table 6.1: Exponents of the XY, mean-field, and chiral universality class exponents as determined by experiment and theory.

	α	β	γ	ν
Experiment				
Wang *et al.* (1991)	0.39(9)			
Deutschmann *et al.* (1992)	0.40(5)			
Mason *et al.* (1989)		0.21(2)	1.01(8)	0.54(3)
Ajiro *et al.* (1988)		0.25(1)		
Kadowaki *et al.* (1988)			1.10(5)	0.57(3)
Gaulin *et al.* (1989)		0.24(2)		
Plakhty *et al.* (2000)		0.21(1)		
Simulation & theory				
XY	−0.02	0.345	1.32	0.67
mean-field tricritical	0.5	0.25	1	0.5
Nagano *et al.* (2019)	0.44(3)	0.26(2)	1.03(5)	0.52(1)
Kawamura (1998)	0.34(6)	0.253(10)	1.13(5)	0.54(2)
Plumer and Mailhot (1994)	0.46(10)	0.24(2)	1.03(4)	0.50(1)
Boubcheur *et al.* (1996)	0.46(10)	0.25(2)	1.15(5)	0.48(2)
Pelissetto *et al.* (2001)			1.13(5)	0.57(3)

The transition in CsMnBr$_3$ occurs at a tetracritical point, shown in Fig. 6.2, (Gaulin *et al.*, 1989; Mason *et al.*, 1990) and this is proposed as an alternate explanation for the observed unusual critical behavior. Nevertheless, the complete set of critical exponents, including chiral exponents in Table 6.1, supports chiral universality. In addition to the usual set of critical points, exponents associated with the chiral ordering can be defined along with appropriate scaling relations. For example, the total chirality has the associated exponent β_c, the susceptibility the exponent γ_c, and the crossover exponent ϕ_c. Several scaling exponents relationships hold, which include the chiral exponents (Kawamura, 1992; Plakhty *et al.*, 2000), including

$$\alpha + 2\beta_c + \gamma_c = 2, \tag{6.1}$$

$$\gamma_c = 2\phi_c - (2 - \alpha), \tag{6.2}$$

$$\beta_c = 3\nu - \phi_c, \tag{6.3}$$

$$\nu_c = \nu, \tag{6.4}$$

$$\gamma_c = 2\phi_c - 3\nu, \tag{6.5}$$

where the exponents with the subscript are associated with the chiral order and the others are the conventional exponents that obey the usual scaling relationships.

Neutron scattering studies indicated critical behavior distinct from the unfrustrated $d = 3$ XY behavior, as summarized in Table 6.1. Examples of the power law behavior for the conventional exponents β, κ, and χ are shown in Figs. 6.3 and 6.4. Overall, the results for β, γ and ν are fairly consistent among the various experiments listed in Table 6.1. The specific heat critical behavior is shown versus T in Fig. 6.5. A semi-log plot of

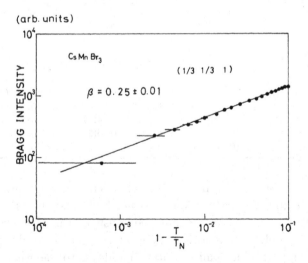

Fig. 6.3: The critical behavior of the order parameter and exponent β for CsMnBr$_3$ (Ajiro *et al.*, 1988). Reproduced with permission from the Physical Society of Japan.

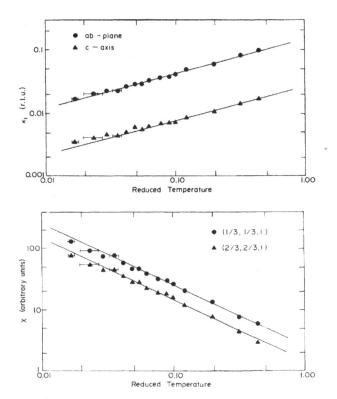

Fig. 6.4: The critical behavior of the inverse fluctuation correlation length κ and the staggered susceptibility χ for CsMnBr$_3$ (Mason *et al.*, 1989). Reproduced with permission from American Physical Society.

the specific heat data is shown in Fig. 6.6 along with a fit to the data. The data showed no signs of significant crossover behavior when fit over different ranges of reduced temperature or when possible crossover terms were added to the fitting function. The value of α is shown in Table 6.1. A similar value (Deutschmann *et al.*, 1992) is also shown in the table. Although not shown in the table, values for the specific heat amplitude ratio were obtained from fits to the data, but the error estimates for the amplitude ratios are large. However, none of the results are consistent with the mean-field value $A^+/A^- = 0$.

It is difficult to directly access the chiral ordering using conventional techniques. However, polarized neutron scattering studies determined the chiral critical exponents listed in Table 6.2 and these are generally consistent with calculated values listed.

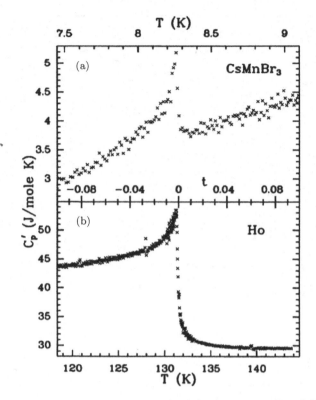

Fig. 6.5: Specific heat versus T of CsMnBr$_3$ (upper) and holmium (lower) (Wang *et al.*, 1992).

Overall, the experiments, theory and simulations support the existence of a chiral universality class distinct from the normal $d = 3$ XY and mean-field tricritical universal behaviors. Once again, it is the combination of a variety of complementary techniques that allows that conclusion to be made.

6.2 Examples of Other Chiral Systems

6.2.1 *Holmium*

Another case where one might expect to see chiral universality is in helical metals such as holmium. The state of the measured critical behavior has been reviewed (Kawamura, 1992) and it is clear that a consensus regarding the exponents does not exist and therefore will not be covered here. An example of the specific heat (Wang *et al.*, 1991) is shown in Figs. 6.5 and 6.6.

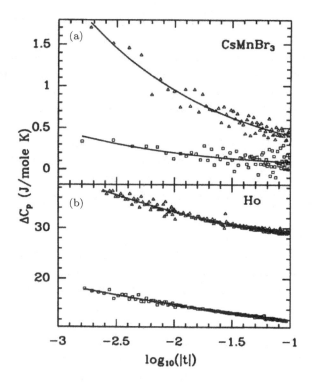

Fig. 6.6: Semi-log plot of the data in Fig. 6.5 (Wang *et al.*, 1992).

Table 6.2: Chiral exponents as determined by experiment and theory.

	β_c	γ_c	ϕ_c
Experiment			
Plakhty *et al.* (2000)	0.44(2)	0.84(7)	1.29(7)
Plakhty *et al.* (2006)		0.85(3)	1.292(23)
Simulation & Theory			
Pelissetto *et al.* (2001)	0.28(10)		1.43(4)
Nagano *et al.* (2019)	0.40(3)	0.77(6)	
Kawamura (1992)	0.45(2)	0.77(5)	1.22(6)
Plumer and Mailhot (1994)	0.38(2)	0.90(9)	1.28(10)

The contrast with the specific heat of CsMnBr$_3$, also shown in the figures, is striking. Attempts at fitting the specific heat of holmium were far from convincing, in contrast to the case of CsMnBr$_3$, and the resulting exponents were inconsistent with the chiral university class. One possibility for the difficulties encountered in the characterization of the critical behavior of holmium lies in the source of the interactions that lead to the helical spin ordering, the RKKY interaction. The interaction determines the helical pitch, and it is relatively long-ranged, unlike the typical near-neighbor interactions in the insulating magnets. The asymptotic universal critical behavior is observed when the correlation length for fluctuations becomes much larger than the longest interaction. While that is easily achieved in insulators as the transition is approached, it is likely that the asymptotic behavior is never adequately realized in metals such as holmium. The inconsistencies noted in the review are consistent with strong crossover effects for temperatures far outside the asymptotic critical behavior.

6.2.2 VF_2

In the review (Kawamura, 1998) of chiral critical behaviors, it was suggested that VF_2, which shows chiral ordering and is an insulator, could be used to overcome the shortcomings of the RKKY interaction in holmium. This suggestion motivated neutron scattering measurements (Ye *et al.*, 2007). The body-centered tetragonal lattice has interactions that result in a rotation of the XY anisotropy spins by approximately $96°$ with respect to their neighbors in the plane perpendicular to the c-axis, which results in a right- or left-handed chiral state. The temperature dependence of the Bragg intensity, which is proportional to the square of the order parameter, was measured and is shown in Fig. 6.7. Power law fits to data yielded an exponent $\beta = 0.175(15)$. This is clearly incompatible with the normal $d = 3$ XY model (Table 6.1) critical behavior for which $\beta = 0.345$, but it is also somewhat smaller than the value for the XY stacked triangular lattice chiral system.

6.3   Ising Stacked Triangular Lattice

As with other stacked triangular antiferromagnets, the interactions in $CsCoBr_3$ are strong along the Co chains and weak in the triangular planes. In the case of $CsCoBr_3$, the ordered spins are perpendicular to the triangular planes. With this Ising anisotropy, there are three sublattices which cannot all order simultaneously and satisfy the antiferromagnetic interactions. Two sublattices order antiparallel to each other and the third remains paramagnetic.

The specific heat has been measured (Wang *et al.*, 1994) using a traditional heat pulse technique; the birefringence signal is too weak to be of

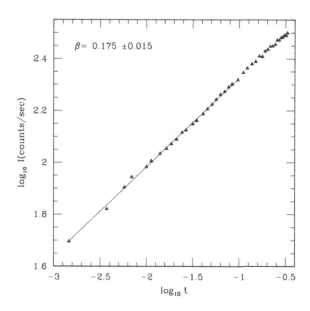

Fig. 6.7: The critical behavior of the order parameter of the chiral phase transition in VF$_2$ (Ye *et al.*, 2007).

use in this case. The specific heat versus temperature (Wang *et al.*, 1991), shown in Fig. 6.8, has a background signal that is large compared to the magnetic component, making the analysis difficult. The data were initially interpreted by fitting to a power law behavior plus smooth background terms,

$$C = A^{\pm}|t|^{-\alpha} + D + Et + Ft^2 + Gt^3, \qquad (6.6)$$

where $t = (T - T_N)/T_N$, with the resulting exponents $\alpha = -0.025(4)$ and $A^+/A^- = 1.07(2)$. The fits, shown in Fig. 6.9, are of excellent quality. Encouragingly, this fits reasonably well with the XY universality class, which was believed to govern this transition. Some experimental results for other exponents and some Monte Carlo simulations suggested this to be the case. Earlier neutron scattering measurements of the order parameter (Farkas *et al.*, 1991) did not fit with the XY universality class however. To resolve this conflict, high-resolution neutron scattering studies were made (Mao *et al.*, 2002) with results shown in Figs. 6.10 and 6.11. Fits yielded exponents for β, γ, and ν that are more consistent with tricritical behavior, and Monte Carlo simulations suggested tricritical behavior with mean-field-like exponents as well. The problem is that the tricritical exponent for the specific heat is $\alpha = 1/2$, which is clearly incompatible with the measured

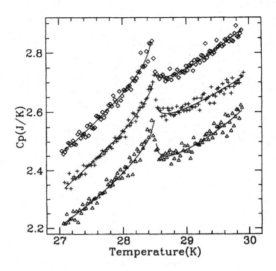

Fig. 6.8: The specific heat behavior of CsCoBr$_3$ versus temperature in three separate measurements, with vertical offsets for clarity (Wang *et al.*, 1991).

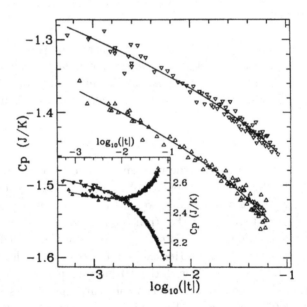

Fig. 6.9: A semi-log plot of the same data shown in Fig. 6.8 with the background subtracted (Wang *et al.*, 1991). The inset shows the data without the background subtraction.

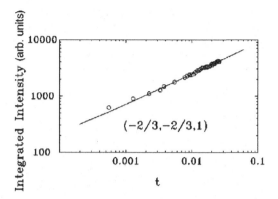

Fig. 6.10: The tricritical fits of neutron scattering results for CsCoBr₃ (Mao *et al.*, 2002) for the temperature dependence of the order parameter with the exponent β. Reproduced with permission from American Physical Society.

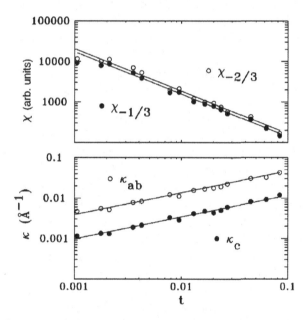

Fig. 6.11: The tricritical fits of neutron scattering results for CsCoBr₃ (Mao *et al.*, 2002) for the temperature dependence of the staggered susceptibility and inverse correlation length with exponents γ, and ν, respectively. Reproduced with permission from American Physical Society.

Table 6.3: Exponents of the XY and tricritical universality classes. The tricritical exponents are exact. The XY exponents are consistent with most results from theory and simulations listed by Pelissetto and Vicari (2002). Monte Carlo results are from Meloche and Plumer (2007) with $H = 0$ and $H = 0.25$.

	α	β	γ	ν
Tricritical	0.5	0.25	1.0	0.5
XY	−0.02	0.345	1.32	0.67
Farkas *et al.* (1991)		0.22(1)		
Wang *et al.* (1994)	−0.025(4)			
Mao *et al.* (2002)		0.28(2)	1.05(8)	0.54(5)
Monte Carlo $H = 0$		0.35(2)	1.33(5)	0.670(7)
Monte Carlo $H = 0.25$		0.28(3)	1.22(6)	0.56(1)

exponent using Eq. (6.6). A similar situation is found in simulation work, as reviewed by Meloche and Plumer (2007). The expected exponents for the tricritical and XY universality classes, measured exponents CsCoBr$_3$, and recent Monte Carlo results (Meloche and Plumer, 2007) are summarized in Table 6.3.

To resolve the conflicting experimental results, the specific heat data were fit (Mao *et al.*, 2002) to expected tricritical behavior given by

$$C^{\pm} = C_0^{\pm}(1 + 3|t|)^{-1/2}, \qquad (6.7)$$

where $t = (T - T_N)/(T_N - T_{cr})$, T_N is the transition temperature and T_{cr} is the crossover temperature, where $T_N - T_{cr} = 0.012\,\mathrm{K}$. The results are shown in Fig. 6.12 and can be compared to the original fits in Fig. 6.8. The tricritical behavior fits are fairly reasonable descriptions of the data, lending support to the interpretation of all the data as representing tricritical critical behavior. However, the tricritical fits in Fig. 6.12 are not as good as the power law fits in Fig. 6.8. In critical behavior studies, the interpretation of data is dependent on the context of the model used in the analyses. Crossover effects can be more apparent in some critical behavior measurements than in others, and one possibility is that both Eqs. (6.6) and (6.7) are approximations. To fit the specific heat accurately, a fitting function with XY to tricritical crossover incorporated might be more appropriate. However, the weak signal compared to background terms would make this a difficult strategy to implement.

Recent Monte Carlo simulations (Meloche and Plumer, 2007) suggest that there might be a crossover from XY to tricritical behavior as the

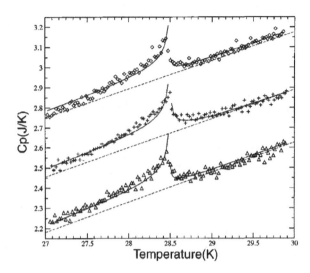

Fig. 6.12: The fit (Mao *et al.*, 2002) of the specific heat versus T data shown in Fig. 6.8 to tricritical behavior described by Eq. (6.7). Reproduced with permission from American Physical Society.

transition is approached. The crossover is attributed to quantum effects associated with short-range order in the chains that can be emulated in the simulation by the introduction of a small applied field. The results of the $H = 0$ simulations in Table 6.3 are consistent with XY critical behavior, whereas the $H = 0.25$ results indicate possible tricritical behavior. Instructive points regarding this analysis are that the model being used can impact the interpretation of the critical behavior data and that it is essential to explore several aspects of the critical behavior in order to arrive at a reliable characterization of the universality class governing a particular phase transition.

Chapter 7

The Unusual Magnetism of LaCoO$_3$: A Thermally Excited Exchange Interaction and Ordering at Twin Interfaces

This chapter introduces an example showing how the characterization of the critical behavior, in this case of the order parameter of LaCoO$_3$, can give indications about the source of magnetism in a crystal. The crystal structure is illustrated in Fig. 7.1. The interactions in LaCoO$_3$ crystals are antiferromagnetic at temperatures above $100\,K$, as determined from fits to the Curie–Weiss approximation (Eqs. (2.21) and (2.22)) (Belanger et al., 2016; Durand et al., 2013). Yet, there is no transition to long-range bulk antiferromagnetic order at low temperatures. As described below, this is a consequence of the temperature dependence of the lattice parameters, as shown in Fig. 7.2.

A new class of critical behavior was predicted (Binder and Hohenberg, 1972, 1974) for transitions that are supported by bulk magnetic ordering, but taking place at a surface. The order-parameter critical exponent was predicted to be profoundly distinct from that of the bulk transition in the same crystal with the order parameter exponent $\beta \approx 0.6 - 0.7$. This is much larger than most phase transitions (see Table 2.1, for example) and is even higher than the mean-field transition exponent value $\beta = 1/2$. Although the unusual exponent was calculated, it was pointed out by the authors that experimental studies would be greatly impeded because the surface magnetism would be drowned out by the much larger bulk magnetic order. Recently, it was realized that LCoO$_3$ is a special case that has allowed possible observation of the surface ordering exponent in a bulk crystal.

The magnetism of LaCoO$_3$ has largely eluded a definitive understanding (Jiang et al., 2009; Sundaram et al., 2009). LaCoO$_3$ has strong

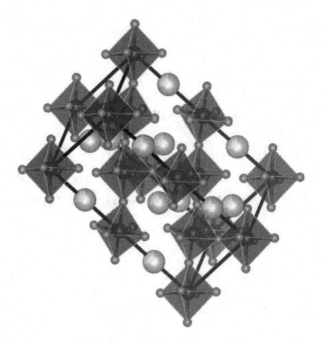

Fig. 7.1: The structure of LaCoO$_3$ with the oxygen octahedra surrounding each Co ion highlighted. The La ions are light-colored. The octahedra share a corner oxygen with neighboring octahedra with a tilt equal to the Co–O–Co bond angle which is near 163°. The magnetic moments are centered on the Co sites and the interactionsare antiferromagnetic above $T = 100$ K, but shut off as T decreases towards 40 K.

antiferromagnetic interactions above 40 K, as evidenced by fits to the Curie–Weiss law, as shown in Fig. 7.3 (see Chapter 2). According to the Curie–Weiss fits (Durand *et al.*, 2013), LaCoO$_3$ should order around $T = 100$ K, and yet there is almost no ordering in large crystals. The literature attempting to explain this is vast, but after decades of research there is little consensus as to why this is except that the magnetic state is slightly above the nonmagnetic ground state and can be activated thermally (Belanger *et al.*, 2016; Lee and Harmon, 2013). Experimentally, the antiferromagnetic interactions weaken as the temperature decreases towards 40 K and, near $T = 100$ K, the interactions are not sufficiently strong to allow long-range order except at defects.

For LaCoO$_3$ grown in bulk particle and nanoparticle forms, a transition involving a weak net moment can occur with T_N between 70 and 90 K. However, for small particles of approximate size 20 nm and with few impurities, the transition is not observed, as can be seen in Fig. 7.3 (upper

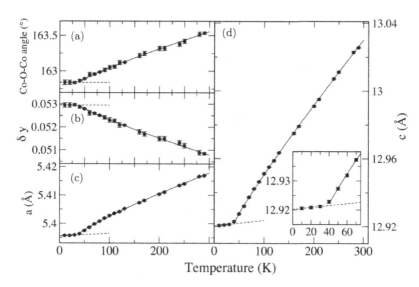

Fig. 7.2: The temperature dependence of the lattice parameters of LaCoO$_3$ measured using neutron powder diffraction (Durand *et al.*, 2013). The lattice parameters increase with T above 40 K, including the Co–O–Co angle, but vary little at lower temperatures. The Co–O–Co angle decreases to 163° at $T = 40$ K; the antiferromagnetic interactions concomitantly decrease and are suppressed below $T = 40$ K. This is explained by the thermal activation of the magnetic interactions (Lee and Harmon, 2013).

figure). Powders C and D consist of particles near 20 nm in size but D has significant Co$_3$O$_4$ impurities while C has few. This indicates that, in the powders, impurities are a source of the net moment. For large particles of size greater than 100 nm, the core region is similar to bulk LaCoO$_3$, but the surfaces have larger lattice parameters. This results in strain that can also induce a transition. When a net ordered moment is observed, the order-parameter exponent is large, with $\beta \approx 0.6$. This suggests surface ordering in the stressed material surrounding defects or near the particle surface. Large crystals have few impurity defects and the surface area of the crystals is insufficient to produce a significant net moment. The bulk of the large crystals does not order antiferromagnetically. Yet, a very weak net moment is observed (Yan *et al.*, 2004) and that the moment grows as defects are introduced by stressing and damaging the crystal. The measured order parameter exponent suggests that the small net moment is associated with surfaces.

Structure measurements correlate the loss of the antiferromagnetic interaction strength with the temperature dependence of the lattice parameters.

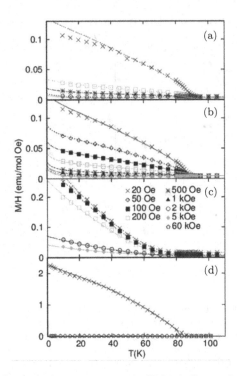

Fig. 7.3: The weak magnetic moment versus T for bulk powder and nanoparticle LaCoO$_3$. The particles have average sizes of > 500 nm (a), 100–400 nm (b), 18 nm (c), and 20 nm (d). Sample (a) and (b) are large enough to have core regions similar to a bulk crystal, whereas the surfaces have larger lattice parameters, leading to strain near the particle surfaces. The (c) and (d) particles are too small to have a core that is different from the surface. However, although the sizes of (c) and (d) are similar, sample (c) has 11% Co$_3$O$_4$ and (d) has 28% Co$_3$O$_4$. When strain near the surface of the particle, or when Co$_3$O$_4$ impurities are present in sufficient numbers, a transition near 80 K occurs with a net moment. The order-parameter exponent $\beta \approx 0.6$ obtained from fits to the net moment is unusually high and suggests surface ordering (Durand *et al.*, 2015).

In particular, when the Co–O–Co bond angle, which describes the tilting angle between corner sharing oxygen octahedra surrounding each Co ion, is less than 163°, the interactions are shut off (Belanger *et al.*, 2016; Durand *et al.*, 2013; Lee and Harmon, 2013). The angle approaches that angle as T decreases to 40 K as the thermally excited sites in the magnetic state depopulate, as shown in Fig. 7.2. Hence, while antiferromagnetism is indicated by the Curie-Weiss behavior at higher temperatures, as illustrated in Fig. 7.4, the number of interactions decrease as the temperature decreases, so long-range order never develops.

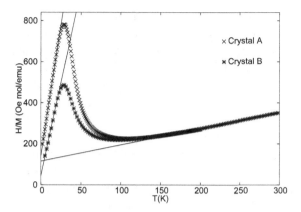

Fig. 7.4: The inverse of the weak magnetic moment versus T for bulk crystals of LaCoO₃ near to and below the transition versus T. The crystal A has more defects. The straight line well above the transition temperature is a fit to Curie–Weiss behavior and suggests strong antiferromagnetic interactions for $T > 150\,\text{K}$ (Belanger *et al.*, 2016). Similar behavior was observed in nanoparticles (Durand *et al.*, 2013).

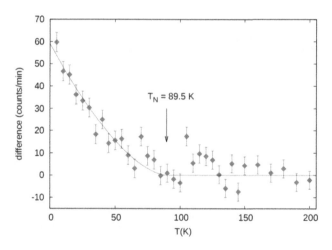

Fig. 7.5: Spin-flip polarized neutron scattering Bragg intensity, which is proportional to the square of the net moment, versus T for a crystal of LaCoO₃, showing the unusual shape below the transition temperature that is consistent with the large value $\beta > 0.5$. The moment in the crystal is extremely weak.

Even with the absence of antiferromagnetic ordering in most of the crystal, a weak net moment is observed, as noted above. Polarized neutron scattering measurements (Kaminsky *et al.*, 2018) shown in Fig. 7.5 indicate that $\beta \approx 0.6$, so surfaces are again implicated. The Bragg scattering intensity is proportional to the square of the magnetic ordering. Normally, because $\beta \leq 1/2$, the Bragg scattering rises steeply at the transition, but in this case the curvature is reversed and it is difficult to identify the transition temperature using the Bragg intensity; if $\beta > 1/2$, this is the expected behavior. The surface area of the bulk crystal is too small to account for the net moment. Likewise, impurities, especially Co_3O_4, are insufficient in number to produce the net moment. However, twinning interfaces are ubiquitous in $LaCoO_3$. It was shown through modeling (Kaminsky *et al.*, 2018) that twins cause some Co–O–Co bonds to remain above 163°, allowing the moments near the twin interface to order and create a net moment. This model is not precisely the model of surface ordering proposed (Binder and Hohenberg, 1972, 1974) because the bulk does not order. However, the bulk of the crystal is very close to ordering and can still assists the ordering at the twin interface surfaces. It was the unusual order parameter value $\beta > 1/2$ that helped to identify the twinning in $LaCoO_3$ as the source of the small net moment.

Chapter 8

Conclusions and Outstanding Questions

8.1 Overall Summary of the Results of the Experiments

Numerous examples of critical behavior experiments were presented in this book to show how the universal properties of second-order phase transitions can be determined. Pure systems without frustration or crossover effects can be measured to a high degree of precision. However, crossover effects can be difficult to characterize. Experiments must be analyzed keeping the possibility of the influence of crossover effects in mind. Even more challenging are the effects of concentration gradients in diluted and mixed magnetic systems. Nevertheless, understanding their consequences and how to minimize them can lead to good characterizations of the intrinsic critical behavior.

For the most part, the experimental results are in good agreement with theory and simulations. When discrepancies are uncovered, new physical understanding can be the result of experimentalists and theorists working towards agreement. One area in which large disagreements were prevalent in the early years is the $d = 3$ random-field Ising model. Not only were some of the experiments and theory at odds, experimentalists were often drawing contradictory conclusions based on the data being collected. The differing interpretations of experiments could sometimes be traced to the manner in which the effects of concentration gradients were accounted for. A major issue in the early theory and experimental interpretations had to do with whether or not a $d = 3$ phase transition actually takes place with random fields applied. The consensus now is overwhelming that a new critical behavior is present with random fields.

Great effort was made to understand concentration gradients and to produce crystals in which they are minimized. That effort allowed much more detailed characterizations of the underlying universal behaviors in diluted magnetic crystals.

Throughout this book, experiments were presented that have served as archetypal examples of universality classes that govern the properties near magnetic phase transitions. Classic experiments in pure systems without frustration showed the precision possible in the experiments and the excellent agreement with theory and computer simulations. In pure systems with frustration, the symmetry changes led to new universality classes. In some systems, predicted chiral ordering was observed.

A large part of the book dealt with the effects of random-site dilution. In particular, for zero applied fields, the predicted change in the universality class behavior for the case of the $d = 3$ Ising system, where the pure system specific heat exponent is $\alpha > 0$, was verified through experiments. When a field is applied to dilute anisotropic antiferromagnets, random-field Ising behavior can be studied. For $d = 2$, the experiments showed an equilibrium destruction of the phase transition once the field is applied, consistent with theory and the prediction of $d = 2$ being the lower critical dimension. The $d = 3$ case has proven to be difficult to characterize in experiments. The predicted activated dynamics were verified in experiments. The extremely slow dynamic behavior prevents truly static measurements. Nevertheless, a great deal of progress has been made in the experiments, though there are areas where the experiments and theory have yet to form a definitive characterization of the critical behavior. The predicted $d = 3$ random-field transition was observed and the universal critical behaviors are still being studied.

For the well-studied fluoride systems used in the critical behavior experiments, additional topics were covered. The phase diagrams have been studied in some detail. While the multicritical points of the pure systems have been characterized in detail, the effects of dilution are more difficult to characterize. The emergence of the spin-glass-like behavior and other nonequilibrium phenomena have been extensively studied. Ising cluster excitations were studied in dilute anisotropic antiferromagnets and detailed studies of fracton excitations in dilute isotropic systems were covered. At low temperatures, the dynamic behavior of domains in zero applied field has been characterized in the dilute anisotropic antiferromagnets.

Part of the success of the experimental investigations covered in the book comes from the extensive characterizations of the single crystals used in the studies. The experiments, along with the work done in theory and computer simulations, provide a solid base for investigations of phase transitions in other materials such as thin films and nanoparticles where the structures are more complicated.

One topic that was thoroughly discussed was the proportionality between $d(\Delta n)/dT$ and the magnetic specific heat. The question is an important one because it provided an accurate technique that in some cases surpassed the precision of direct specific heat and because it circumvented some of the difficulties with concentration gradients. A great deal of evidence shows that, in the fluorides, the proportionality is extremely accurate. Speculations questioning that proportionality proved to be unwarranted. Some of the remaining issues in bringing theory and experiment together on the characterization of the $d = 3$ random field Ising model are centered on the critical behavior of specific heat, as discussed below.

Another area that proves difficult in the study of dilute magnets, and especially the random-field systems, comes from the extremely slow activated dynamics. Although the consensus of theory and experiments is that the systems are dominated by activated dynamics, so much so that the concept of static, equilibrium measurements becomes clouded; it is still not very clear how to interpret experimental and simulated results in such systems. This is discussed below with respect to the order parameter and specific heat, which are normally considered static measurements.

An interesting theme that occurs throughout many of the studies of phase transition, and particularly systems with quenched randomness is the idea of fractal structures. This is observed in behaviors near the magnetic percolation threshold concentration, near the vacancy percolation threshold concentration, and in the behaviors near the magnetic phase transitions. Particularly important is the role of fractal structures near the $d = 3$ random-field Ising transition and how they interact with the critical behavior.

A final lesson learned from the studies of dilute magnets is that site dilution can affect the experiments in ways that are not clear from scaling arguments. In particular, for $d = 3$ systems, vacancies can form a percolating lattice that greatly weakens the magnetic ordering. This likely affects the dynamic critical behavior near the phase transition.

The critical behavior in pure frustrated systems appears to be well understood in the classic experiments covered in this book. The chiral universality is well established and is distinct from universality classes of the unfrustrated magnetic phase transitions. The observation of the chiral exponents complements the normal exponents. The Ising triangular lattice provides an additional example of new universal behavior associated with frustrated transitions.

It also appears that the dynamics of domain walls at low temperatures is fairly well understood. The studies in the fluorides and chlorides serve as models for low temperature dynamics in more complicated systems.

8.2 Some Open Questions About Equilibrium in the Random-Field Ising Model

The most difficult phase transition to study, and the one with still outstanding fundamental problems to solve, is the $d = 3$ random-field Ising model. A large section of the book was used to detail the experimental results so that the areas still to be investigated will have a comprehensive base upon which to build. The study of the model systems are made difficult by the extremely slow activated dynamics that turn even what are normally static measurements into dynamic ones. Another aspect of the $d = 3$ random-field measurements is the role of vacancy percolation for $x < v_v$, where x_v is the vacancy percolation threshold concentration. For $x < v_v$, there is a strong propensity for the system to form metastable domain structures. Even for $x > x_v$, the system shows nonequilibrium behavior near the transition and there are lingering questions regarding the critical behavior that need to be addressed.

From the order parameter experiments with $x > x_v$, it is clear that a critical exponent β can be measured in $Fe_x Zn_{1-x} F_2$ with an accuracy rivaling that of experiments using FeF_2, but only if the temperature increases. Even the slightest reversal of temperature ruins the power law behavior. These measurements have been done using synchrotron radiation that probes length scales larger than those that neutron scattering can probe. A natural conclusion might be that the system cannot achieve long-range equilibrium order just below the transition upon cooling, but can probe equilibrium critical behavior upon heating from low temperatures where long-range order is already established. However, is the ZFC procedure showing us equilibrium behavior? The difficulty in achieving equilibrium critical behavior has been known for a long time (see, for example Ogielski, 1986). A $d = 3$ Monte Carlo simulation (Barber and Belanger, 2000) with parameters approximating the physical properties of FeF_2 and 2×128^3 spins each randomly visited up to 5000 times, was done to see how closely it would mimic the order-parameter experiments. Using the ZFC procedure, the random-exchange and random-field order-parameter critical behavior is remarkably similar in appearance to the experimental results, as shown in the upper panel of Fig. 8.1. To obtain the results shown in Fig. 8.1,

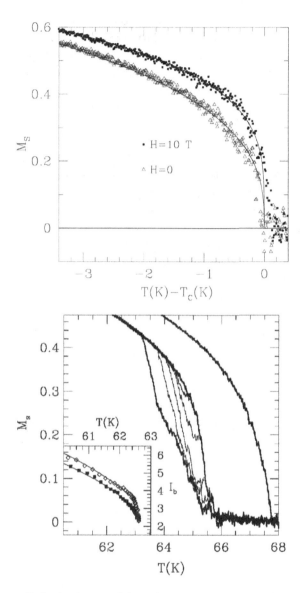

Fig. 8.1: Monte Carlo simulations of the order-parameter critical behavior of the $d = 3$ random-exchange and random-field Ising model. The ZFC behaviors are shown in the upper panel (Barber and Belanger, 2000) and the ZFC-FC hysteresis is shown in the lower panel (Shelton *et al.*, 2004). In the lower panel, the right-most curve is $H = 0$. The smoother curve for $H > 0$ is ZFC. The jagged curves are FC.

minimal effort was made, intentionally, to ensure the system was in equilibrium. Another Monte Carlo simulation (Shelton *et al.*, 2004), shown in the lower panel of Fig. 8.1, exhibits ZFC-FC hysteresis bearing a remarkable resemblance to the experiments. It would be interesting to have a better understanding of the behavior of the experimental hysteresis that is so easily mimicked by a Monte Carlo simulation where great lengths to ensure equilibrium are not taken. Most Monte Carlo simulations are done on small sizes and great effort is made to ensure equilibrium. However, is that what the experiments are measuring? Although a long-range magnetic structure exists after ZFC, is this the equilibrium long-range order of the random-field Ising model? The specific heat appears sharp under ZFC, but does it represent the asymptotic critical behavior of fluctuations on the equilibrium magnetic structure?

In the same spirit, a similar Monte Carlo simulation (Barber and Belanger, 2000) was done, again without a great effort to ensure equilibrium, to see how easy it was to mimic the $d = 3$ random-field specific heat behavior upon ZFC. The results are shown in Fig. 8.2. Surprisingly, the results resemble the $H = 0$ random-exchange and the ZFC random-field

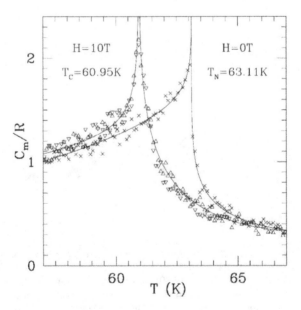

Fig. 8.2: A Monte Carlo simulation of the specific heat of the $d = 3$ random-exchange and random-field Ising model (Barber and Belanger, 2000).

Ising experimental data, with an asymmetric peak for $H = 0$ and a symmetric peak for $H > 10\,\mathrm{T}$. Monte Carlo studies on smaller lattices, where equilibrium has been meticulously ensured, have not come to a consensus regarding the asymptotic random-field Ising critical behavior. Is the experimental critical behavior the same as the equilibrium Monte Carlo studies?

8.3 Future Work

Hopefully, progress will continue to be made, especially on the random-field Ising behavior. In particular, questions about the scattering line shapes are outstanding as is the nature of the specific heat critical behavior. Slow dynamics in the diluted anisotropic crystals have been explored for $x < x_v$, but similar studies at higher concentrations, especially for $H > 0$, could be quite interesting.

Bibliography

Aharony, A. (1978a). Spin-flop multicritical points in systems with random fields and in spin glasses, *Phys. Rev. B* **18**, p. 3328, 10.1103/PhysRevB.18.3328.

Aharony, A. (1978b). Tricritical points in systems with random fields, *Phys. Rev. B*, p. 3318 10.1103/PhysRevB.18.3318.

Aharony, A. (1986). Crossover from random exchange to random field critical behavior, *Europhys. Lett.* **1**, p. 617, 10.1209/0295-5075/1/12/002.

Aharony, A., Imry, Y., and k. Ma, S. (1976). Lowering of dimensionality in phase transitions with random fields, *Phys. Rev. Lett.* **37**, p. 1364, 10.1103/PhysRevLett.37.1364.

Ahlers, G., Kornblit, A., and Salamon, M. B. (1974). Heat capacity of FeF_2 near the antiferromagnetic transition, *Phys. Rev. B* **9**, p. 3932, 10.1103/PhysRevB.9.3932.

Ahrens, B., Xiao, J., Hartmann, A. K., and Katzgraber, H. G. (2013). Diluted antiferromagnets in a field seem to be in a different universality class than the random-field Ising model, *Phys. Rev. B* **88**, p. 174408, 10.1103/PhysRevB.88.174408.

Aizenman, M. and Wehr, J. (1989). Rounding of 1st-order phase-transitions in systems with quenched disorder, *Phys. Rev. Lett.* **62**, p. 2503, 10.1103/PhysRevLett.62.2503.

Ajiro, Y., Nakashima, T., Unno, Y., Kadowaki, H., Mekata, M., and Achiwa, N. (1988). New critical exponent β of the XY antiferromagnet on stacked triangular lattice, $CsMnBr_3$, *J. Phys. Soc. Jnp.* **57**, p. 2648, 10.1143/JPSJ.57.2648.

Álvarez, G., Aso, N., Belanger, D. P., Durand, A. M., Martín-Mayor, V., Motoya, K., and Muro, Y. (2012). Neutron scattering experiments and simulations near the percolation threshold of $Fe_xZn_{1-x}F_2$, *Phys. Rev. B* **86**, p. 024416, 10.1103/PhysRevB.86.024416.

Ballesteros, H. G., Fernández, L. A., Martín-Mayor, V., Sudupe, A. M., Parisi, G., and Ruiz-Lorenzo, J. J. (1998). Critical exponents of the three dimensional diluted Ising model, *Phys. Rev. B* **58**, p. 2740, 10.1103/PhysRevB.58.2740.

Barak, J., Jaccarino, V., and Rezende, S. M. (1978). The magnetic anisotropy of MnF_2 at 0K, *J. Magn. Magn. Mater.* **9**, p. 323, 10.1016/0304-8853(78)90087-2.

Barber, W. C. (1999). Monte Carlo program.

Barber, W. C. and Belanger, D. P. (2000). The random field critical concentration in dilute antiferromagnets, *J. Appl. Phys.* **87**, p. 7049, 10.1063/1.372927.

Barber, W. C., Ye, F., Belanger, D. P., and Fernandez-Baca, J. A. (2004). Magnetic vacancy percolation in dilute antiferromagnets, *Phys. Rev. B* **69**, p. 024409, 10.1103/PhysRevB.69.024409.

Barbosa, P. H. R., Raposo, E. P., and Coutinho-Filho, M. D. (2000). Monte Carlo studies of the spin-glass-line phase in $Fe_{0.25}Zn_{0.75}F_2$, *J. Appl. Phys.* **87**, p. 6531, 10.1063/1.372760.

Barbosa, P. H. R., Raposo, E. P., and Coutinho-Filho, M. D. (2003). Microscopic description of an Ising spin glass near the percolation threshold, *Phys. Rev. Lett.* **91**, p. 197207, 10.1103/PhysRevLett.91.197207.

Barbosa, P. H. R., Raposo, E. P., and Coutinho-Filho, M. D. (2005). Monte Carlo study of the random-exchange-Ising-model behavior of a diluted antiferromagnet: $Fe_{0.48}Zn_{0.52}F_2$, *Phys. Rev. B* **72**, p. 092401, 10.1103/PhysRevB.72.092401.

Barrett, P. H. (1986). Static and dynamic critical phenomena in $Fe_{1-x}Zn_xF_2$, *Phys. Rev. B* **34**, p. 3513, 10.1103/PhysRevB.34.3513.

Baur, W. H. (1958). Uber die verfeinerung der kristallstrukturbestimmung einiger vertreter des rutiltyps. II. Die difluoride von Mn, Fe, Co, Ni und Zn, *Acta Cryst.* **11**, p. 488, 10.1107/S0365110X58001353.

Becerra, C. C., Paduan-Filho, A., Fries, T., Shapira, Y., Gabas, M., Campo, J., and Palacio, F. (1995). Low-field remanent magnetization in site-dilted easy-axis antiferromagnets: a universal behavior, *J. Magn. Magn. Mater.* **140–144**, p. 1475, 10.1016/0304-8853(94)01355-1.

Belanger, D., King, A. R., Ferreira, I. B., and Jaccarino, V. (1988a). Concentration inhomogeneities in random magnets II. Effects on critical phenomema studies, *Phys. Rev. B* **37**, p. 226, 10.1103/PhysRevB.37.226.

Belanger, D. P. (1981). *Linear birefringence studies of magnetic critical phenomena*, Ph.D. thesis, University of California, Santa Barbara.

Belanger, D. P., Farago, B., Jaccarino, V., King, A. R., Lartigue, C., and Mezei, F. (1988b). Random exchange Ising model dynamics: $Fe_{0.46}Zn_{0.54}F_2$, *J. de Phys.* **49**, pp. C8–1229, 10.1051/jphyscol:19888557.

Belanger, D. P., Jaccarino, V., King, A. R., and Nicklow, R. M. (1987). Neutron scattering observations of extreme critical slowing down in a $d = 3$ random field system, *Phys. Rev. Lett.* **59**, p. 930, 10.1103/PhysRevLett.59.930.

Belanger, D. P., Keiber, T., Bridges, F., Durand, A. M., Mehta, A., Zheng, H., Mitchell, J. F., and Borzenets, V. (2016). Structure and magnetism in $LaCoO_3$, *J. Phys.: Condens. Matter* **28**, p. 025602, 10.1088/0953-8984/28/2/025602.

Belanger, D. P., King, A. R., and Jaccarino, V. (1982a). Critical behavior in random anisotropic antiferromagnets, *J. Appl. Phys.* **53**, p. 2704, 10.1063/1.330938.

Belanger, D. P., King, A. R., and Jaccarino, V. (1982b). Random field effects in diluted anisotropic antiferromagnets, *J. Appl. Phys.* **53**, p. 2702, 10.1063/1.330937.

Belanger, D. P., King, A. R., and Jaccarino, V. (1982c). Random field effects on critical behavior of diluted Ising antiferromagnets, *Phys. Rev. Lett.* **48**, p. 1050, 10.1103/PhysRevLett.48.1050.

Belanger, D. P., King, A. R., and Jaccarino, V. (1984). Temperature dependence of the optical birefringence of MnF_2, MgF_2 and ZnF_2, *Phys. Rev. B* **29**, p. 2636, 10.1103/PhysRevB.29.2636.

Belanger, D. P., King, A. R., and Jaccarino, V. (1985a). Neutron scattering study of hysteresis near T_C in a $d = 3$ random field system, *Sol. St. Comm.* **54**, p. 79, 10.1016/0038-1098(85)91038-5.

Belanger, D. P., King, A. R., and Jaccarino, V. (1985b). Random field critical behavior of a $d = 3$ ising system: Neutron scattering studies of $Fe_{0.6}Zn_{0.4}F_2$, *Phys. Rev. B* **31**, p. 4538, 10.1103/PhysRevB.31.4538.

Belanger, D. P., King, A. R., and Jaccarino, V. (1985c). Scaling of the metastability boundary of a $d = 2$ random field Ising system, *Phys. Rev. Lett.* **54**, p. 577, 10.1103/PhysRevLett.54.577.

Belanger, D. P., King, A. R., and Jaccarino, V. (1986). Crossover from random exchange to random field critical behavior in $Fe_xZn_{1-x}F_2$, *Phys. Rev. B* **34**, p. 452, 10.1103/PhysRevB.34.452.

Belanger, D. P., King, A. R., Jaccarino, V., and Cardy, J. L. (1983a). Random-field critical behavior of a $D = 3$ Ising system, *Phys. Rev. B* **28**, p. 2522, 10.1103/PhysRevB.28.2522.

Belanger, D. P., Kleemann, W., and Montenegro, F. C. (1996). X-ray and neutron scattering, magnetization, and heat capacity study of the 3D random field Ising model — Comment, *Phys. Rev. Lett.* **77**, p. 2341, 10.1103/PhysRevLett.77.2341.

Belanger, D. P., Murray, W. E., Montenegro, F. C., King, A. R., Jaccarino, V., and Erwin, R. W. (1991). Neutron-scattering study of an Ising antiferromagnet above the percolation threshold: Random-field-induced spinglass-like behavior in $Fe_{0.31}Zn_{0.69}F_2$, *Phys. Rev. B* **44**, p. 2161, 10.1103/PhysRevB.44.2161.

Belanger, D. P., Nordblad, P., King, A. R., Jaccarino, V., Lundgren, L., and Beckman, O. (1983b). Critical behavior in anisotropic antiferromagnets, *J. Mang. Magn. Mater.* **31–34**, p. 1095, 10.1016/0304-8853(83)90813-2.

Belanger, D. P., Rezende, S. M., King, A. R., and Jaccarino, V. (1985d). Hysteresis, metastability, and time dependence in $d = 2$ and $d = 3$ random-field Ising systems, *J. Appl. Phys.* **57**, p. 3294, 10.1063/1.335126.

Belanger, D. P. and Yoshizawa, H. (1987). Neutron scattering and the critical behavior of the three dimensional Ising antiferromagnet FeF_2, *Phys. Rev. B* **35**, p. 4823, 10.1103/PhysRevB.35.4823.

Belanger, D. P. and Yoshizawa, H. (1993). Neutron-scattering study of the random-exchange Ising system $Fe_xZn_{1-x}F_2$ near the percolation threshold, *Phys. Rev. B* **47**, p. 5051, 10.1103/PhysRevB.47.5051.

Belanger, D. P. and Young, A. P. (1991). The random-field Ising model, *J. Magn. Magn. Mater.* **100**, p. 272, 10.1016/0304-8853(91)90825-U.

Bensamka, F., Bertrand, D., Fert, A. R., Jehanno, G., and Nasser, J. A. (1987). Mössbauer spectroscopy of the spin-glass compounds $Fe_xMg_{1-x}Cl_2$, *J. Phys.* **48**, p. 1041, 10.1051/jphys:019870048060104100.

Bertrand, D., Redoules, J. P., Ferre, J., Pommier, J., and Souletie, J. (1988). Dynamic evidence for a transition at $T_C = 0$ in the Ising anisotropic spin-glass $Fe_{0.3}Mg_{0.7}Cl_2$, *Euro. Phys. Lett.* **5**, p. 271, 10.1209/0295-5075/5/3/015.

Binder, K. and Hohenberg, P. C. (1972). Phase-transitions and static spin correlations in Ising models with free surfaces, *Phys. Rev. B* **6**, p. 3461, 10.1103/PhysRevB.6.3461.

Binder, K. and Hohenberg, P. C. (1974). Surface effects on magnetic phase-transitions, *Phys. Rev. B* **9**, p. 2194, 10.1103/PhysRevB.9.2194.

Binder, K. and Young, A. P. (1986). Spin glasses: Experimental facts, theoretical concepts, and open questions, *Rev. Mod. Phys.* **58**, p. 801, 10.1103/RevModPhys.58.801.

Binek, C., Kleemann, W., and Belanger, D. P. (1998). Crossover from pure Ising to random-exchange dominated behavior of the two-dimensional antiferromagnet $Rb_2Co_{1-x}Mg_xF_4$, *Phys. Rev. B* **57**, p. 7791, 10.1103/PhysRevB.57.7791.

Birgeneau, R. J., Aharony, A., Cowley, R. A., Hill, J. P., Pelcovits, R. A., Shirane, G., and Thurston, T. R. (1991). Effects of random fields on bicritical phase diagrams in two and three dimensions, *Physica* **177**, p. 58, 10.1016/0378-4371(91)90134-X.

Birgeneau, R. J., Cowley, R. A., Shirane, G., and Yoshizawa, H. (1985). Temporal phase transition in the three-dimensional random-field Ising model, *Phys. Rev. Lett.* **54**, p. 2147, 10.1103/PhysRevLett.54.2147.

Birgeneau, R. J., Cowley, R. A., Shirane, G., Yoshizawa, H., Belanger, D. P., King, A. R., and Jaccarino, V. (1983a). Critical-behavior of a site-diluted 3-dimensional Ising magnet, *Phys. Rev. B* **27**, p. 6747, 10.1103/PhysRevB.27.6747.

Birgeneau, R. J., Feng, Q., Harris, Q. J., Hill, J. P., and Ramirez, A. P. (1996). Birgenau *et al.* reply, *Phys. Rev. Lett.* **77**, p. 2342, 10.1103/PhysRevLett.77.2342.

Birgeneau, R. J., Shirane, G., Blume, M., and Koehler, W. C. (1974). Tricritical point in $FeCl_2$, *Phys. Rev. Lett.* **33**, p. 1098, 10.1103/PhysRevLett.33.1098.

Birgeneau, R. J., Yoshizawa, H., Cowley, R. A., Shirane, G., and Ikeda, H. (1983b). Random-field effects in the diluted two-dimensional Ising antiferromagnet $Rb_2Co_{0.7}Mg_{0.3}F_4$, *Phys. Rev. B* **28**, p. 1438, 10.1103/PhysRevB.28.1438.

Boo, W. O. J. and Stout, J. W. (1976). Heat capacity and entropy of MnF_2 from 10 to 300K. Evaluation of the contributions associated with magnetic ordering, *J. Chem. Phys.* **65**, p. 3929, 10.1063/1.432885.

Boubcheur, E. H., Loison, D., and Diep, H. T. (1996). Phase diagram of XY antiferromagnetic stacked triangular lattices, *Phys. Rev. B* **54**, p. 4165, 10.1103/PhysRevB.54.4165.

Breed, D. J., Gilijamse, K., and Miedema, A. R. (1969). Magnetic properties of K_2CoF_4 and Rb_2CoF_4: Two-dimensional Ising antiferromagnets, *Physica* **45**, p. 205, 10.1016/0031-8914(69)90073-1.

Bricmont, J. and Kupiainen, A. (1987). Lower critical dimension for the random-field Ising-model, *Phys. Rev. Lett.* **59**, p. 1829, 10.1103/PhysRevLett.59. 1829.

Bruce, A. D. (1981). The theory of super-lattice critical scattering, *J. Phys. C: Sol. State Phys.* **14**, p. 193, 10.1088/0022-3719/14/3/004.

Calabrese, P., Pelissetto, A., and Vicari, E. (2003). Crossover from random-exchange to random-field critical behavior in Ising models, *Phys. Rev. B* **68**, p. 092409, 10.1103/PhysRevB.68.092409.

Cardy, J. L. (1984). Random-field effects in site-disordered Ising antiferromagnets, *Phys. Rev. B* **29**, p. 505, 10.1103/PhysRevB.29.505.

Chamberlin, R. V. (1994). Mesoscopic model for the primary response of magnetic materials, *J. Appl. Phys.* **76**, p. 6401, 10.1063/1.358278.

Chen, K. and Landau, D. P. (1993). Spin dynamics in the diluted three-dimensional classical Heisenberg antiferromagnet, *J. Appl. Phys.* **73**, p. 5645, 10.1063/1.353626.

Chirwa, M., Lundgren, L., Nordblad, P., and Beckman, O. (1980). Magnetic specific heat of FeF_2 near T_N, *J. Magn. Magn. Mater.* **15–18**, p. 457, 10. 1016/0304-8853(80)91129-4.

Collins, M. F. (1989). *Magnetic Critical Scattering* (Oxford University Press, New York).

Collins, M. F. and Petrenko, O. A. (1997). Triangular antiferromagnets, *Can. J. Phys.* **75**, p. 605, 10.1139/cjp-75-9-605.

Cowley, R. A., Aharony, A., Birgeneau, R. J., Pelcovits, R. A., and Thurston, T. R. (1993). The bicritical phase diagram of two-dimensional antiferromagnets with and without random fields, *Z. Phys. B* **93**, p. 5, 10.1007/BF01308802.

Cowley, R. A., Hagen, M., and Belanger, D. P. (1984a). An experimental study of the critical fluctuations in a two-dimensional Ising model, *J. Phys. C* **17**, p. 3763, 10.1088/0022-3719/17/21/010.

Cowley, R. A., Shirane, G., Yoshizawa, H., Uemura, Y. J., and Birgeneau, R. J. (1989). The effects of random fields on the critical and bicritical behavior of $Mn_xZn_{1-x}F_2$, *Z. Phys. B* **75**, p. 303, 10.1007/BF01321818.

Cowley, R. A., Yoshizawa, H., Shirane, G., and Birgeneau, R. J. (1984b). Random field effects in the three dimensional Ising magnet: $Fe_xZn_{1-x}F_2$, *Z. Phys. B* **58**, p. 15, 10.1007/BF01469433.

Cowley, R. A., Yoshizawa, H., Shirane, G., Hagen, M., and Birgeneau, R. J. (1984c). Random fields and the weakly anisotropic Ising model: $Mn_xZn_{1-x}F_2$, *Phys. Rev. B* **30**, p. 6650, 10.1103/PhysRevB.30.6650.

da Cunha, J. B. M., de Araújo, J. H., Amaral, L., Vasquez, A., Moro, J. T., Montenegro, F. C., Rezende, S. M., and Coutinho-filho, M. D. (1990). Mössbauer effect measurements on the spin-glass $Fe_{0.25}Zn_{0.75}F_2$, *Hyp. Int.* **54**, p. 489, 10.1007/BF02396077.

de Almeida, J. R. L. and Thouless, D. J. (1978). Stability of the Sherrington-Kirkpatrick solution of a spin glass model, *J. Phys. A: Math. Gen.* **11**, p. 983, 10.1088/0305-4470/11/5/028.

de Araújo, J. H., da Cunha, J. B. M., Vasquez, A., Amaral, L., Moro, J. T., Montenegro, F. C., Rezende, S. M., and Coutinho-Filho, M. D. (1991). Mössbauer study of spin-glass $Fe_xZn_{1-x}F_2$ system, *Hyp. Int.* **67**, p. 507, 10.1007/BF02398192.

de Lima, K. A. P., Brito, J. B., Barbosa, P. H. R., Raposo, E. P., and Coutinho-Filho, M. D. (2012). Magnetic field effect on the fractal cluster spin-glass phase of an Ising antiferromagnet near the first-neighbor percolation threshold: $Fe_{0.25}Zn_{0.75}F_2$, *Phys. Rev. B* **85**, p. 064416, 10.1103/PhysRevB.85.064416.

Delfino, G. (1998). Universal amplitude ratios in the two-dimensional Ising model, *Phys. Lett. B* **419**, p. 291, 10.1016/S0370-2693(97)01457-3.

Deutschmann, R., v. Lohneysen, H., Wosnitza, J., Kremer, R. K., and Visser, D. (1992). Critical behavior in the specific heat of an antiferromagnet with chiral symmetry, *Europhys. Lett.* **17**, p. 637, 10.1209/0295-5075/17/7/011.

Djurberg, C., Mattsson, J., and Nordblad, P. (1994). Remanent magnetization in the diluted Ising antiferromagnet $Fe_{0.6}Zn_{0.4}F_2$, *J. Appl. Phys.* **75**, p. 5541, 10.1063/1.355682.

Dotsenko, V. S. and Dotsenko, V. S. (1982). Critical-behavior of the 2D Ising-model with impurity bonds, *J. Phys. C: Sol. St. Phys.* **15**, p. 495, 10.1088/0022-3719/15/3/015.

Dow, K. E. and Belanger, D. P. (1989). The specific heat of $Fe_{0.46}Zn_{0.54}F_2$, *Phys. Rev. B* **39**, p. 4418, 10.1103/PhysRevB.39.4418.

Durand, A. M., Belanger, D. P., Booth, C. H., Ye, F., Chi, S., Fernandez-Baca, J. A., and Bhat, M. (2013). Magnetism and phase transitions in $LaCoO_3$, *J. Phys.: Condens. Matter* **25**, p. 382203, 10.1088/0953-8984/25/38/382203.

Durand, A. M., Hamil, T. J., Belanger, D. P., Chi, S., Ye, F., Fernandez-Baca, J. A., Abdollahian, Y., and Booth, C. H. (2015). The effects of Co_3O_4 on the structure and unusual magnetism of $LaCoO_3$, *J. Phys.: Condens. Matter* **27**, p. 126001, 10.1088/0953-8984/27/12/126001.

Esser, J., Nowak, U., and Usadel, K. D. (1997). Exact ground-state properties of disordered Ising systems, *Phys. Rev. B* **55**, p. 5866, 10.1103/PhysRevB.55.5866.

Farkas, A., Gaulin, B. D., Tun, Z., and Briat, B. (1991). Magnetic phase-transitions in $CsCoBr_3$, *J. Appl. Phys.* **69**, p. 6167, 10.1063/1.348794.

Fernandez, L. A., Martín-Mayor, V., and Yllanes, D. (2011). Critical behavior of the dilute antiferromagnet in a magnetic field, *Phys. Rev. B* **84**, p. 100408, 10.1103/PhysRevB.84.100408.

Ferre, J. and Gehring, G. A. (1984). Linear optical birefringence of magnetic crystals, *Rep. Prog. Phys.* **47**, p. 513, 10.1088/0034-4885/47/5/002.

Ferre, J., Jamet, J. P., and Kleemann, W. (1982). Critical behavior of the linear birefringence in 3d-Heisenberg antiferromagnets: $RbMnF_3$, $KNiF_3$, and $KCoF_3$, *Sol. St. Comm.* **44**, p. 485, 10.1016/0038-1098(82)90129-6.

Ferreira, I. B., Cardy, J. L., King, A. R., and Jaccarino, V. (1991a). Concentration scaling of the specific heat in the $d = 3$ random exchange Ising model system $Fe_x Zn_{1-x} F_2$, *J. Appl. Phys.* **69**, p. 5075, 10.1063/1.348127.

Ferreira, I. B., King, A. R., and Jaccarino, V. (1991b). Concentration dependence of the random field crossover scaling in $Fe_x Zn_{1-x} F_2$, *J. Appl. Phys.* **69**, p. 5246, 10.1063/1.348093.

Ferreira, I. B., King, A. R., and Jaccarino, V. (1991c). Concentration dependence of the random-field-crossover scaling in $Fe_x Zn_{1-x} F_2$, *Phys. Rev. B* **43**, p. 10797, 10.1103/PhysRevB.43.10797.

Ferreira, I. B., King, A. R., Jaccarino, V., Cardy, J. L., and Guggenheim, H. J. (1983). Random-field-induced destruction of the phase transition of a diluted two-dimensional Ising antiferromagnet: $Rb_{0.85} Mg_{0.15} F_4$, *Phys. Rev. B* **28**, p. 5192, 10.1103/PhysRevB.28.5192.

Fisher, D. S. (1986). Scaling and critical slowing down in random-field Ising systems, *Phys. Rev. Lett.* **56**, p. 416, 10.1103/PhysRevLett.56.416.

Fisher, M. E. (1962). Relation between specific heat and susceptibility of an antiferromagnet, *Phil. Mag.* **7**, p. 1731, 10.1080/14786436208213705.

Fisher, M. E. (1967). The theory of equilibrium critical phenomena, *Rep. Prog. Phys.* **30**, p. 615, 10.1088/0034-4885/30/2/306.

Fisher, M. E. and Burford, R. J. (1967). Theory of critical-point scattering and correlations. I. Ising model, *Phys. Rev.* **156**, p. 583, 10.1103/PhysRev.156.583.

Fishman, S. and Aharony, A. (1979). Random field effects in disordered anisotropic anti-ferromagnets, *J. Phys. C: Sol. St. Phys.* **12**, p. L729, 10.1088/0022-3719/12/18/006.

Fytas, N. G. and Malakis, A. (2011). Scaling and self-averaging in the three-dimensional random-field Ising model, *Eur. Phys. J. B* **79**, p. 13, 10.1140/epjb/e2010-10404-6.

Fytas, N. G. and Martín-Mayor, V. (2013). Universality in the three-dimensional random-field Ising model, *Phys. Rev. Lett.* **110**, p. 227201, 10.1103/PhysRevLett.110.227201.

Fytas, N. G., Martín-Mayor, V., Picco, M., and Sourlas, N. (2018). Review of recent developments in the random-field ising model, *J. Stat. Phys.* **172**, p. 665, 10.1007/s10955-018-1955-7.

Fytas, N. G., Theodorakis, P. E., and Hartmann, A. K. (2016). Revisiting the scaling of the specific heat of the three-dimensional random-field Ising model, *Eur. Phys. J. B* **89**, p. 200, 10.1140/epjb/e2016-70364-3.

Gaulin, B. D., Mason, T. E., Collins, M. F., and Larese, J. Z. (1989). Tetracritical behavior of $CsMnBr_3$, *Phys. Rev. Lett.* **62**, p. 1380, 10.1103/PhysRevLett.62.1380.

Gehring, G. A. (1977). Observation of critical indexes of primary and secondary order parameters using birefringence, *J. Phys. C — Sol. St. Phys.* **10**, p. 531, 10.1088/0022-3719/10/4/010.

Glaser, A., Jones, A. C., and Duxbury, P. M. (2005). Domain states in the zero-temperature diluted antiferromagnet in an applied field, *Phys. Rev. B* **71**, p. 174423, 10.1103/PhysRevB.71.174423.

Griffiths, R. B. (1969). Nonanalytic behavior above critical point in a random Ising ferromagnet, *Phys. Rev. Lett.* **23**, p. 17, 10.1103/PhysRevLett.23.17.

Hagen, M., Cowley, R. A., Satija, S. K., Yoshizawa, H., Shirane, G., Birgeneau, R. J., and Guggenheim, H. J. (1983). Random fields and three-dimensional Ising models: $Co_xZn_{1-x}F_2$, *Phys. Rev. B* **28**, p. 2602, 10.1103/PhysRevB. 28.2602.

Hagen, M. and Paul, D. M. (1984). An experimental test of 2-scale-factor universality, *J. Phys. C: Sol. St. Phys.* **17**, p. 5605, 10.1088/0022-3719/17/31/ 016.

Han, S.-J. and Belanger, D. P. (1992). Relaxation of the remanent magnetization of dilute anisotropic antiferromagnets, *Phys. Rev. B* **46**, p. 2926, 10.1103/ PhysRevB.46.2926.

Han, S.-J., Belanger, D. P., Kleemann, W., and Nowak, U. (1992). Relaxation of the excess magnetization of random field induced metastable domains $Fe_{0.47}Zn_{0.53}F_2$, *Phys. Rev. B* **45**, p. 9728, 10.1103/PhysRevB.45.9728.

Harris, A. B. (1974). Effect of random defects on the critical behavior of Ising models, *J. Phys. C: Sol. St. Phys.* **7**, p. 1671, 10.1088/0022-3719/7/9/009.

Hartmann, A. K. and Young, A. P. (2001). Specific-heat exponent of random-field systems via ground-state calculations, *Phys. Rev. B* **64**, p. 214419, 10. 1103/PhysRevB.64.214419.

Hasenbusch, M., Pelissetto, A., and Vicari, E. (2007). Relaxational dynamics in 3D randomly diluted Ising models, *J. Stat. Mech.* , p. P1100910.1088/ 1742-5468/2007/11/P11009.

Hill, J. P., Feng, Q., Harris, Q. J., Birgeneau, R. J., Ramirez, A. P., and Cassanho, A. (1997). Phase-transitition behavior in the random-field antiferromagnet $Fe_{0.5}Zn_{0.5}F_2$, *Phys. Rev. B* **55**, p. 356, 10.1103/PhysRevB.55. 356.

Hohenberg, P. C., Aharony, A., Halperin, B. I., and Siggia, E. D. (1976). Two-scale universality and the renormalization group, *Phys. Rev. B* **13**, p. 2986, 10.1103/PhysRevB.13.2986.

Hohenberg, P. C. and Halperin, B. I. (1977). Theory of dynamic critical phenomena, *Rev. Mod. Phys.* **49**, p. 435, 10.1103/RevModPhys.49.435.

Hohenemser, C., Rosov, N., and Kleinhammes, A. (1989). Critical phenomena studied via nuclear techniques, *Hyper. Inter.* **49**, p. 267, 10.1007/ BF02405146.

Hutchings, M. T., Rainford, B. D., and Guggenheim, H. J. (1970). Spin waves in antiferromagnetic FeF_2, *J. Phys. C. Sol. St. Phys.* **3**, p. 307, 10.1088/ 0022-3719/3/2/013.

Ikeda, H. (1994). Spin dynamics in the two-dimensional percolating Ising antiferromagnet, *Physica A* **204**, p. 328, 10.1016/0378-4371(94)90434-0.

Ikeda, H., Fernandez-Baca, J. A., Nicklow, R. M., Takahashi, M., and Iwasa, K. (1994). Fracton excitations in a dilute Heisenberg antiferromagnet near the percolation threshold: $RbMn_{0.39}Mg_{0.61}F_3$, *J. Phys.: Cond. Matter* **6**, p. 10543, 10.1088/0953-8984/6/48/015.

Ikeda, H., Hatta, I., Ikushima, A., and Hirakawa, K. (1975). Observation of a symmetric logarithmic singularity in the specific heat of K_2CoF_4, *J. Phys. Soc. Jpn.* **39**, p. 827, 10.1143/JPSJ.39.827.

Ikeda, H., Hatta, I., and Tanaka, M. (1976). Critical heat-capacities of 2-dimensional Ising-like antiferromagnets K_2CoF_4 and Rb_2CoF_4, *J. Phys. Soc. Jpn.* **40**, p. 334, 10.1143/JPSJ.40.334.

Ikeda, H. and Hutchings, M. (1978). Spin-wave excitations in a 2-dimensional Ising-like anti-ferromagnet, Rb_2CoF_4, *J. Phys. C: Sol. St. Phys.* **11**, p. L529, 10.1088/0022-3719/11/13/007.

Ikeda, H., Itoh, S., and Adams, M. A. (1995a). Anomalous spin diffusion in a two-dimensional percolating Ising antiferromagnet, *Phys. Rev. Lett.* **75**, p. 4440, 10.1103/PhysRevLett.75.4440.

Ikeda, H., Itoh, S., Adams, M. A., and Fernandez-Baca, J. A. (1998a). Crossover from homogeneous to fractal excitations in the near-percolating Heisenberg antiferromagnet $RbMn_{0.39}Mg_{0.61}F_3$, *J. Phys. Soc. Jpn.* **67**, p. 3376, 10.1143/JPSJ.67.3376.

Ikeda, H., Iwasa, K., Fernandez-Baca, J. A., and Nicklow, R. M. (1995b). Spin correlations in percolating networks with fractal geometry, *Physica B* **213 & 214**, p. 146, 10.1016/0921-4526(95)00087-P.

Ikeda, H., Okamura, N., Kato, K., and Ikushima, A. (1978). Experimental-observation of crossover phenomena in specific heat of MnF_2, *J. Phys. C — Sol. St. Phys.* **11**, p. L231, 10.1088/0022-3719/11/7/001.

Ikeda, H., Suzuki, M., and Hutchings, M. T. (1979). Neutron scattering investigation of static critical phenomena in the two-dimensional antiferromagnets: $Rb_2Co_cMg_{1-c}F_4$, *J. Phys. Soc. Jpn.* **46**, p. 1153, 10.1143/JPSJ.46.1153.

Ikeda, H., Takahashi, M., Fernandez-Baca, J. A., and Nicklow, R. M. (1998b). Neutron study of fracton excitations in percolating antiferromagnets, *J. Magn. Magn. Mater.* **177–181**, p. 139, 10.1016/S0304-8853(97)00334-X.

Imbrie, J. Z. (1984). Lower critical dimension of the random-field Ising-model, *Phys. Rev. Lett.* **53**, p. 1747, 10.1103/PhysRevLett.53.1747.

Imry, Y. and Ma, S. (1975). Random-field instability of ordered state of continuous symmetry, *Phys. Rev. Lett.* **35**, p. 1399, 10.1103/PhysRevLett.35.1399.

Itoh, S., Ikeda, H., Yoshizawa, H., Harris, M. J., and Steigenberger, U. (1998). Spin dynamics in two-dimensional percolating Heisenberg antiferromagnets, *J. Phys. Soc. Jpn.* **67**, p. 3610, 10.1143/JPSJ.67.3610.

Itoh, S., Kajimoto, R., Adams, M. A., Bull, M. J., Iwasa, K., Aso, N., Yoshizawa, H., and Takeuchi, T. (2007). Fractal dimension in percolating Heisenberg antiferromagnets, *J. Magn. Magn. Mater.* **310**, p. 1549, 10.1016/j.jmmm.2006.10.560.

Itoh, S., Nakayama, T., and Adams, M. A. (2011). Antiferromagnetic fractons in diluted Heisenberg systems $RbMn_{0.4}Mg_{0.6}F_3$ and $Rb_{0.598}Mg_{0.402}F_4$, *J. Phys. Soc. Jpn.* **80**, p. 104704, 10.1143/JPSJ.80.104704.

Itoh, S., Nakayama, T., and Adams, M. A. (2012). Two-dimensional antiferromagnetic fractons in $Rb_2Mn_{0.598}Mg_{0.402}F_4$, *J. Phys. Conf. Series* **400**, p. 032030, 10.1088/1742-6596/400/3/032030.

Itoh, S., Nakayama, T., Kajimoto, R., and Adams, M. A. (2009). Single-length-scaling analysis for antiferromagnetic fractons in dilute Heisenberg system $RbMn_{0.4}Mg_{0.6}F_3$, *J. Phys. Soc. Jpn.* **78**, p. 013707, 10.1143/JPSJ.78.013707.

Jaccarino, V. and King, A. R. (1990). Random exchange and random field Ising systems: Static and dynamic critical behavior, *Phys. A* **163**, p. 291, 10.1016/0378-4371(90)90338-S.

Jaccarino, V., King, A. R., Motokawa, M., Sakakibara, T., and Date, M. (1983). Temperature dependence of FeF_2 spin flop field, *J. Magn. Magn. Mater.* **31–34**, p. 1117, 10.1016/0304-8853(83)90822-3.

Jahn, I. R. (1973). Linear magnetic birefringence in the antiferromagnetic iron group difluorides, *Phys. Stat. Sol. (b)* **57**, p. 681, 10.1002/pssb.2220570225.

Jahn, I. R., Merkel, J. B., Gehring, G. A., and Becker, P. J. (1977). Critical behavior of the magnetic birefringence in $(Mn,Zn)F_2$, *Physica* **89B**, p. 177, 10.1016/0378-4363(77)90075-4.

Jiang, Y., Bridges, F., Sundaram, N., Belanger, D. P., Anderson, I. E., Mitchell, J. F., and Zheng, H. (2009). Study of the local distortions of the perovskite system $La_{1-x}Sr_xCo_3$ $0 \leq x \leq 0.35$ using the extended x-ray absorption fine structure technique, *Phys. Rev. B* **80**, p. 144423, 10.1103/PhysRevB.80.144423.

Jonason, K., Djurberg, C., Nordblad, P., and Belanger, D. P. (1997). Dynamics of the antiferromagnetic random-exchange Ising system $Fe_xZn_{1-x}F_2$ near the percolation threshold, *Phys. Rev. B* **56**, p. 5404, 10.1103/PhysRevB.56.5404.

Kadowaki, H., Shapiro, S. M., Inami, T., and Ajiro, Y. (1988). New universality class of antiferromagnetic phase-transition in $CsMnBr_3$, *J. Phys. Soc. JPN.* **57**, p. 2640, 10.1143/JPSJ.57.2640.

Kaminsky, G. M., Belanger, D. P., Ye, F., Fernandez-Baca, J. A., Wang, J., Matsuda, M., and Yan, J.-Q. (2018). Origin of the net magnetic moment in $LaCoO_3$, *Phys. Rev. B* **97**, p. 024418, 10.1103/PhysRevB.97.024418.

Kato, T., Iio, K., Hoshino, T., Mitsui, T., and Tanaka, H. (1992). Birefringence study on structural and magnetic phase transitions in $RbMnBr_3$, *J. Phys. Soc. Jpn.* **61**, p. 275, 10.1143/JPSJ.61.275.

Kawamura, H. (1986). Phase transition of the three-dimensional XY antiferromagnet on the layered-triangular lattice, *J. Phys. Soc. Jpn.* **55**, p. 2095, 10.1143/JPSJ.55.2095.

Kawamura, H. (1992). Monte Carlo study of chiral criticality — XY and Heisenberg stacked-triangular antiferromagnets, *J. Phys. Soc. Jpn.* **61**, p. 1299, 10.1143/JPSJ.61.1299.

Kawamura, H. (1998). Universality of phase transitions of frustrated antiferromagnets, *J. Phys.: Condens. Matter* **10**, p. 4707, 10.1088/0953-8984/10/22/004.

King, A. R., Belanger, D. P., Nordblad, P., and Jaccarino, V. (1984). Capacitance measurement of magnetic specific heat, *J. Appl. Phys.* **55**, p. 2410, 10.1063/1.333678.

King, A. R., Ferreira, I. B., Jaccarino, V., and Belanger, D. P. (1988). Concentration inhomogeneities in random magnets I. Characterization using optical birefringence, *Phys. Rev. B* **37**, p. 219, 10.1103/PhysRevB.37.219.

King, A. R., Jaccarino, V., Belanger, D. P., and Rezende, S. M. (1985). Scaling of the equilibrium boundary of three-dimensional random-field Ising-model systems, *Phys. Rev. B* **32**, p. 503, 10.1103/PhysRevB.32.503.

King, A. R., Jaccarino, V., Sakakibara, T., Motokawa, M., and Date, M. (1983). $H - T$ phase diagram of randomly diluted FeF_2, *Magn. Magn. Mater.* **31–34**, p. 1119, 10.1016/0304-8853(83)90823-5.

King, A. R., Mydosh, J. A., and Jaccarino, V. (1986). AC susceptibilty study of the $d = 3$ random-field critical dynamics, *Phys. Rev. Lett.* **56**, p. 2525, 10.1103/PhysRevLett.56.2525.

King, A. R. and Rohrer, H. (1979). Spin-Flop bicritical point in MnF_2, *Phys. Rev. B* **19**, p. 5864, 10.1103/PhysRevB.19.5864.

Kleemann, W., Jakobs, C., Binek, C., and Belanger, D. P. (1998). Kinetics of random-field-induced domains in the two-dimensional Ising antiferromagnet $Rb_2Co_{0.85}Mg_{0.15}F_4$, *J. Magn. Magn. Mat.* **177–181**, p. 209, 10.1016/S0304-8853(97)00671-9.

Kleemann, W., King, A. R., and Jaccarino, V. (1986). Critical behavior of the magnetization of a $d = 3$ random-field Ising system, *Phys. Rev. B* **34**, p. 479, 10.1103/PhysRevB.34.479.

Kornblit, A. and Ahlers, G. (1973). Heat capacity of $RbMnF_3$ near the antiferromagnetic transition temperature, *Phys. Rev. B* **8**, p. 5163, 10.1103/PhysRevB.8.5163.

Kosterlitz, J. M. and Thouless, D. J. (1973). Ordering, metastability and phase-transitions in 2 dimensional systems, *J. Phys. C — Sol. St. Phys.* **6**, p. 1181, 10.1088/0022-3719/6/7/010.

Kushauer, J. and Kleemann, W. (1995). Critical behaviour of the linear and non-linear magnetic susceptibilities of $FeCl_2$, *J. Phys. Condens. Matter* **7**, p. L1, 10.1088/0953-8984/7/1/001.

Kushauer, J., Kleemann, W., Mattsson, J., and Nordblad, P. (1994). Crossover from logarithmically relaxing to piezomagnetically frozen magnetic remanence in low-field-cooled $Fe_{0.47}Zn_{0.53}F_2$, *Phys. Rev. B* **49**, p. 6346, 10.1103/PhysRevB.49.6346.

Lederman, M., Hammann, J., and Orbach, R. (1990). Net spontaneous magnetisation in the dilute Ising antiferromagnet $Fe_{0.46}Zn_{0.54}F_2$, *Physica B* **165 & 166**, p. 179.

Lederman, M., Selinger, J. V., Bruinsma, R., Hammann, J., and Orbach, R. (1992). Low-temperature dynamics of a diluted Ising antiferromagnet, *Phys. Rev. Lett.* **68**, p. 2086, 10.1103/PhysRevLett.68.2086.

Lederman, M., Selinger, J. V., Bruinsma, R., Orbach, R., and Hammann, J. (1993). Dynamics of the diluted Ising antiferromagnet $Fe_{0.46}Zn_{0.54}F_2$ in the (H,T) plane, *Phys. Rev. B* **48**, p. 3810, 10.1103/PhysRevB.48.3810.

Lee, Y. and Harmon, B. N. (2013). Rhombohedral distortion effects on electronic structure of $LaCoO_3$, *J. Appl. Phys.* **113**, p. 17E145, 10.1063/1.4798350.

Leitão, U. A. and Kleemann, W. (1987). Critical-behavior and metastability of the magnetization in the random-field Ising system $Fe_{0.7}Mg_{0.3}Cl_2$, *Phys. Rev. B* **35**, p. 8696, 10.1103/PhysRevB.35.8696.

Leitão, U. A. and Kleemann, W. (1988). Crossover from random exchange to random field critical behavior: A nonlinear susceptibility study of $Fe_{0.7}Mg_{0.3}Cl_2$, *Euro. Phys. Lett.* **5**, p. 529, 10.1209/0295-5075/5/6/009.

Leitão, U. A., Kleemann, W., and Ferreira, I. (1988a). Scaling and metastability of the random-field system $Fe_{0.7}Mg_{0.3}Cl_2$, *J. Phys. Coll.* **49** (**C8**), pp. C8-1217, 10.1051/jphyscol:19888551.

Leitão, U. A., Kleemann, W., and Ferreira, I. B. (1988b). Metastability of the uniform magnetization in three-dimensional random-field Ising model systems, I. $Fe_{0.7}Mg_{0.3}Cl_2$, *Phys. Rev. B* **38**, p. 4765, 10.1103/PhysRevB.38.4765.

Lines, M. E. (1967). Comparative studies of magnetism in $KNiF_3$ and K_2NiF_4, *Phys. Rev.* **164**, p. 736, 10.1103/PhysRev.164.736.

Mao, M., Gaulin, B. D., Rogge, R. B., and Tun, Z. (2002). Tricritical behavior in a stacked triangular lattice Ising antiferromagnet $CsCoBr_3$, *Phys. Rev. B* **66**, p. 184432, 10.1103/PhysRevB.66.184432.

Marinelli, M., Mercuri, F., Foglietta, F., and Belanger, D. P. (1996). Effect of spin-system fluctuations on the heat transport in $RbMnF_3$ close to the Neel temperature, *Phys. Rev. B* **54**, p. 4087, 10.1103/PhysRevB.54.4087.

Martín-Mayor, V., Pelissetto, A., and Vicari, E. (2002). Critical structure factor in Ising systems, *Phys. Rev. E* **66**, p. 026112, 10.1103/PhysRevE.66.026112.

Mason, T. E., Collins, M. F., and Gaulin, B. D. (1990). Monte Carlo simulations of $CsMnBr_3$, *J. Appl. Phys.* **67**, p. 5421, 10.1063/1.344576.

Mason, T. E., Gaulin, B. D., and Collins, M. F. (1989). Neutron-scattering measurements of critical exponents in $CsMnBr_3$ — A Z_2XS_1 antiferromagnet, *Phys. Rev. B* **39**, p. 586, 10.1103/PhysRevB.39.586.

Mattsson, J., Djurberg, C., and Nordblad, P. (1994). Determination of the critical exponent β from measurements of a weak spontaneous magnetisation in the 3d Ising antiferromagnet FeF_2, *J. Magn. Magn. Mater.* **136**, pp. L23–L28, 10.1016/0304-8853(94)90440-5.

Mattsson, J., Djurberg, C., and Nordblad, P. (2000). Low-temperature magnetization in dilute Ising antiferromagnets, *Phys. Rev. B* **61**, p. 11274, 10.1103/PhysRevB.61.11274.

Meloche, E. and Plumer, M. L. (2007). Staggered field-induced tricritical behavior in the $S = 1/2$ quasi-one-dimensional Ising antiferromagnet on a stacked triangular lattice: Monte Carlo simulations, *Phys. Rev. B* **76**, p. 174430, 10.1103/PhysRevB.76.174430.

Montenegro, F. C., Belanger, D. P., Slanič, Z., and Fernandez-Baca, J. A. (2000a). $d = 3$ random field behavior near percolation, *J. Appl. Phys.* **87**, p. 6537, 10.1063/1.372762.

Montenegro, F. C., Belanger, D. P., Slanič, Z., and Fernandez-Baca, J. A. (2000b). Ordering in the dilute weakly anisotropic antiferromagnet $Mn_{0.35}Zn_{0.65}F_2$, *Phys. Rev. B* **61**, p. 14681, 10.1103/PhysRevB.61.14681.

Montenegro, F. C., Coutinho-Filho, M. D., and Rezende, S. M. (1988a). Ising critical behavior in spin glasses: $Fe_{0.25}Zn_{0.75}F_2$, *J. Phys. Coll.* **49**, pp. C8-1007, 10.1051/jphyscol:19888458.

Montenegro, F. C., Coutinho-Filho, M. D., and Rezende, S. M. (1989). Ising criticality in spin glasses: $Fe_{0.25}Zn_{0.75}F_2$, *Eurohys. Lett.* **8**, p. 383, 10.1209/0295-5075/8/4/014.

Montenegro, F. C., de Jesus, J. C. O., Machado, F. L. A., Montarroyos, E., and Rezende, S. M. (1992). Critical and equilibrium phase boundaries in $Mn_{0.5}Zn_{0.5}F_2$, *J. Magn. Magn. Mater.* **104–107**, p. 277, 10.1016/0304-8853(92)90797-R.

Montenegro, F. C., de Jesus, J. C. O., and Rosales-Rivera, A. (1994). Random-exchange to random-field crossover breaking in $Mn_{0.35}Zn_{0.65}F_2$, *J. Appl. Phys.* **75**, p. 5520, 10.1063/1.355675.

Montenegro, F. C., King, A. R., Jaccarino, V., Han, S.-J., and Belanger, D. P. (1991). Random field induced spin glass-like behavior in the dilute Ising antiferromagnet $Fe_{0.31}Zn_{0.69}F_2$, *Phys. Rev. B* **44**, p. 2155, 10.1103/PhysRevB.44.2155.

Montenegro, F. C., Lima, K. A., Torikachvili, M. S., and Lacerda, A. H. (1998). Phase diagram of the random-field Ising system $Fe_{0.60}Zn_{0.40}F_2$ at intense fields, *J. Magn. Magn. Mater.* **177**, p. 145, 10.1016/S0304-8853(97)00313-2.

Montenegro, F. C., Lima, K. A., Torikachvili, M. S., and Lacerda, A. H. (1999). Random-field effects and glassy behavior in $Fe_xZn_{1-x}F_2$, *Mater. Sci. Forum* **302–3**, p. 371, 10.4028/www.scientific.net/MSF.302-303.371.

Montenegro, F. C., Lima, K. A., Torikachvili, M. S., and Lacerda, A. H. (2000c). Instability of long-range order in a d=3 random-field Ising model system: $Fe_xZn_{1-x}F_2$, *Braz. J. Phys.* **30**, p. 758, 10.1590/S0103-97332000000400023.

Montenegro, F. C., Rezende, S. M., and Coutinho-Filho, M. D. (1988b). Evidence for a spin-glass behavior in the diluted antiferromagnet $Fe_xZn_{1-x}F_2$, *J. Appl. Phys.* **63**, p. 3755, 10.1063/1.340657.

Montenegro, F. C., Rosales-Rivera, A., de Jesus, J. C. O., Montarroyos, E., and Machado, F. L. A. (1995). Random-field-crossover scaling in $Mn_{0.35}Zn_{0.65}F_2$, *Phys. Rev. B* **51**, p. 5849, 10.1103/PhysRevB.51.5849.

Nagano, Y., Uematsu, K., and Kawamura, H. (2019). Monte Carlo study of the critical properties of noncollinear Heisenberg magnets: $O(3) \times O(2)$ universality class, *Physical Review B* **100**, p. 224430, 10.1103/PhysRevB.100.224430.

Nakayama, T., Yakubo, K., and Orbach, R. L. (1994). Dynamical properties of fractal networks: Scaling, numerical simulations, and physical realizations, *Rev. Mod. Phys.* **66**, p. 381, 10.1103/RevModPhys.66.381.

Nash, A. E., King, A. R., and Jaccarino, V. (1991). Experimental verification of activated critical dynamics in the $d = 3$ random-field Ising model, *Phys. Rev. B* **43**, p. 1272, 10.1103/PhysRevB.43.1272.

Nattermann, T., Shapir, Y., and Vilfan, I. (1990). Interface pinning and dynamics in random systems, *Phys. Rev. B* **42**, p. 8577, 10.1103/PhysRevB.42.8577.

Nattermann, T. and Vilfan, I. (1988). Anomalous relaxation in the random-field Ising-model and related systems, *Phys. Rev. Lett.* **61**, p. 223, 10.1103/PhysRevLett.61.223.

Nikotin, O., Lindgard, P. A., and Dietrich, O. W. (1969). Magnon dispersion relation and exchange interations in MnF$_2$, *J. Phys. C: Sol. St. Phys.* **2**, p. 1168, 10.1088/0022-3719/2/7/309.

Nordblad, P., Belanger, D. P., King, A. R., Jaccarino, V., and Guggenheim, H. J. (1983a). Critical behavior of induced linear birefringence in isotropic antiferromagnets, *J. Magn. Magn. Mater.* **31–34**, p. 1093, 10.1016/0304-8853(83)90812-0.

Nordblad, P., Belanger, D. P., King, A. R., Jaccarino, V., and Ikeda, H. (1983b). Critical behavior of two-dimensional Rb$_2$CoF$_4$ as observed by linear birefringence, *Phys. Rev. B* **28**, p. 278, 10.1103/PhysRevB.28.278.

Nordblad, P., Lundgren, L., Figueroa, E., and Beckman, O. (1981). Specific heat and magnetic susceptibility of MnF$_2$ and Mn$_{0.98}$Fe$_{0.02}$F$_2$ near T_N, *J. Magn. Magn. Mater.* **23**, p. 333, 10.1016/0304-8853(81)90056-1.

Nowak, U., Esser, J., and Usadel, K. D. (1996). Dynamics of domains in diluted antiferromagnets, *Physica A* **232**, p. 40, 10.1016/0378-4371(96)00133-1.

Nowak, U. and Usadel, K. D. (1991). Diluted antiferromagnets in a magnetic field: A fractal-domain state with spin-glass behavior, *Phys. Rev. B* **44**, p. 7426, 10.1103/PhysRevB.44.7426.

Nowak, U. and Usadel, K. D. (1992a). Correlations and fractality in random Ising magnets, *Phys. A* **191**, p. 203, 10.1016/0378-4371(92)90528-X.

Nowak, U. and Usadel, K. D. (1992b). Structure of domains in random Ising magnets, *Phys. Rev. B* **46**, p. 8329, 10.1103/PhysRevB.46.8329.

Ogielski, A. T. (1986). Integer optimization and zero-temperature fixed point in Ising random-field systems, *Phys. Rev. Lett.* **57**, p. 1251, 10.1103/PhysRevLett.57.1251.

Ogielski, A. T. and Huse, D. A. (1986). Critical behavior of the three-dimensional dilute antiferromagnet in a field, *Phys. Rev. Lett.* **56**, p. 1298, 10.1103/PhysRevLett.56.1298.

Oleaga, A., Salazar, A., and Bunkov, Y. M. (2014). 3D-XY critical behavior of CsMnF$_3$ from static and dynamic thermal properties, *J. Phys.: Condens. Matt.* **26**, p. 096001, 10.1088/0953-8984/26/9/096001.

Oleaga, A., Salazar, A., Prabhakaran, D., Cheng, J. G., and Zhou, J. S. (2012). Critical behavior of the paramagnetic to antiferromagnetic transition in orthorhombic and hexagonal phases of RMnO$_3$ (R = Sm, Tb, Dy, Ho, Er, Tm, Yb, Lu, Y), *Phys. Rev. B* **85**, p. 184425, 10.1103/PhysRevB.85.184425.

Onsager, L. (1944). Crystal statistics I A two-dimensional model with an order-disorder transition, *Phys. Rev.* **65**, p. 117, 10.1103/PhysRev.65.117.

Paduani, C., Belanger, D. P., Wang, J., Han, S.-J., and Nicklow, R. M. (1994). Magnetic excitations in the dilute anisotropic antiferromagnet Fe$_x$Zn$_{1-x}$F$_2$, *Phys. Rev. B* **50**, p. 193, 10.1103/PhysRevB.50.193.

Palacio, F., Campo, J., Moron, M. C., Morón, M. C., Paduan-Filho, A., and Becerra, C. C. (1998). Impurity-induced magnetic anomalies in the slightly diluted low anisotropy $A_2Fe_{1-x}In_xCl_5 \cdot H_2O$, (A=K, Rb) antiferromagnets: An overview, *Int. J. of Mod. Phys. B* **12**, p. 1781, 10.1142/S0217979298001010.

Palacio, F., Gabás, M., Campo, J., Becerra, C. C., Paduan-Filho, A., Fries, T., and Shapira, Y. (1994). Remanent Magnetization in diluted low-anisotropy antiferromagnetis at very low magnetic fields: New phenomena with universal behavior, *Phys. Scripta* **T55**, p. 163, 10.1088/0031-8949/1994/T55/028.

Pelcovits, R. A. and Aharony, A. (1985). Structure factor for dilute magnetic systems, *Phys. Rev. B* **31**, p. 350, 10.1103/PhysRevB.31.350.

Pelissetto, A., Rossi, P., and Vicari, E. (2001). Chiral exponents in frustrated spin models with noncollinear order, *Phys. Rev. B* **65**, p. 020403, 10.1103/PhysRevB.65.020403.

Pelissetto, A. and Vicari, E. (2002). Critical phenomena and renormalization-group theory, *Physics Reports* **368**, p. 549, 10.1016/S0370-1573(02)00219-3.

Pfeuty, P., Jasnow, D., and Fisher, M. E. (1974). Crossover scaling functions for exchange anisotropy, *Phys. Rev. B* **10**, p. 2088, 10.1103/PhysRevB.10.2088.

Picco, M. and Sourlas, N. (2015). Diluted antiferromagnetic 3D Ising model in a field, *EPL* **109**, p. 37001, 10.1209/0295-5075/109/37001.

Plakhty, V. P., Kulda, J., Visser, D., Moskvin, E. V., and Wosnitza, J. (2000). Chiral critical exponents of the triangular-lattice antiferromagnet $CsMnBr_3$ as determined by polarized neutron scattering, *Phy. Rev. Lett.* **85**, p. 3942, 10.1103/PhysRevLett.85.3942.

Plakhty, V. P., Wosnitza, J., Kulda, J., Brückel, T., Schweika, W., Visser, D., Gavrilov, S. V., Moskvin, E. V., Kremer, R. I., and Banks, M. G. (2006). Polarized neutron scattering studies of chiral criticality, and new universality classes of phase transitions, *Physica B* **385–386**, p. 288, 10.1016/j.physb.2006.05.019.

Plumer, M. L. and Mailhot, A. (1994). Tricritical behavior of the frustrated XY antiferromagnet, *Phys. Rev. B* **50**, p. 16113, 10.1103/PhysRevB.50.16113.

Ramos, C. A., King, A. R., and Jaccarino, V. (1988a). Determination of the crossover exponent in the random-field system $Mn_xZn_{1-x}F_2$, *Phys. Rev. B* **37**, p. 5483, 10.1103/PhysRevB.37.5483.

Ramos, C. A., King, A. R., Jaccarino, V., and Rezende, S. M. (1988b). Field-induced anomalous dilation in $Fe_xZn_{1-x}F_2$, *J. Phys. Colloques* **49**, pp. C8-1241, 10.1051/jphyscol:19888563.

Rezende, S., Montenegro, F., Coutinho-Filho, M., Becerra, C., and Paduan-Filho, A. (1988). Dynamic scaling in the Ising spin glasses: $Fe_{0.25}Zn_{0.75}F_2$, *J. Phys. Coll.* **49**, pp. C8-1267, 10.1051/jphyscol:19888576.

Rezende, S. M., King, A. R., and Jaccarino, V. (1984). Hysteresis effects in the critical behavior of a $d = 3$ Ising random field system, *J. Appl. Phys.* **55**, p. 2413, 10.1063/1.333679.

Rodriguez, Y. W., Anderson, I. E., Belanger, D. P., Nojiri, H., Ye, F., and Fernandez-Baca, J. A. (2007). Low-temperature excitations in a dilute three-dimensional anisotropic antiferromagnet, *J. Magn. Magn. Mater.* **310**, p. 1546, 10.1016/j.jmmm.2006.10.558.

Rosales-Rivera, A., Ferreira, J. M., and Montenegro, F. C. (2000). Random-field and glassy dynamics in a diluted Ising antiferromagnet: $Fe_{0.42}Zn_{0.58}F_2$, *Euro. Phys. Lett.* **50**, p. 264, 10.1209/epl/i2000-00264-2.

Rosales-Rivera, A., Ferreira, J. M., Montenegro, F. C., and Ramos, C. A. (2001). Magnetic behavior of the diluted antiferromagnet $Mn_{0.39}Zn_{0.61}F_2$ at strong fields, *J. Magn. Magn. Mater.* **226–230**, p. 1343, 10.1016/S0304-8853(00)00848-9.

Rosov, N., Kleinhammes, A., Lidbjork, P., Hohenemser, C., and Eibschutz, M. (1988). Single-crystal Mössbauer measurement of the critical exponent β in thr random-exchange Ising system $Fe_{0.9}Zn_{0.1}F_2$, *Phys. Rev. B* **37**, p. 3265, 10.1103/PhysRevB.37.3265.

Sakon, T., Awaji, S., Motokawa, M., and Belanger, D. P. (2002). Large random-field behavior above the vacancy percolation concentration threshold: $Fe_{0.84}Zn_{0.16}F_2$, *J. Phys. Soc. Jpn.* **71**, p. 411, 10.1143/JPSJ.71.411.

Sartorelli, J. C. (1992). NMR determination of the order-parameter exponent β in $Fe_{0.46}Zn_{0.54}F_2$, *Phys. Rev. B* **45**, p. 10779, 10.1103/PhysRevB.45.10779.

Satooka, J., Katori, H. A., Tobo, A., and Katsumata, K. (1998). Absence of hysteresis in the heat capacity of the three-dimensional random-field Ising model, *Phys. Rev. Lett.* **81**, p. 709, 10.1103/PhysRevLett.81.709.

Satooka, J., Katsumata, K., and Belanger, D. P. (2002). Magnetic excitations in the three-dimensional dilute antiferromagnet $Fe_xZn_{1-x}F_2$, *J. Phys.: Cond. Mat.* **14**, p. 1307, 10.1088/0953-8984/14/6/317.

Schwartz, M. and Soffer, A. (1986). Critical correlation susceptibility relation in random-field systems, *Phys. Rev. B* **33**, p. 2059, 10.1103/PhysRevB.33.2059.

Seppälä, E. T. and Alava, M. J. (2001). Susceptibility and percolation in two-dimensional random field Ising magnets, *Phys. Rev. E* **63**, p. 066109, 10.1103/PhysRevE.63.066109.

Seppälä, E. T., Alava, M. J., and Sillanpaa, I. J. (2004). Domain walls in random field Ising magnets: Wetting, *J. Magn. Magn. Mater.* **272**, p. 1286, 10.1016/j.jmmm.2003.12.071.

Seppälä, E. T., Petaja, V., and Alava, M. J. (1998). Disorder, order, and domain wall roughening in the two-dimensional random field Ising model, *Phys. Rev. E* **58**, p. R5217, 10.1103/PhysRevE.58.R5217.

Seppälä, E. T., Pulkkinen, A. M., and Alava, M. J. (2002). Percolation in three-dimensional random field Ising magnets, *Phys. Rev. B* **66**, p. 144403, 10.1103/PhysRevB.66.144403.

Shang, H. T. and Salamon, M. B. (1980). Test of tricritical scaling in ferrous chloride, *J. Magn. Magn. Mat.* **15–18**, p. 419, 10.1016/0304-8853(80)91113-0.

Shapir, Y. (1985). Nonperturbative critical behavior of random-field systems, *Phys. Rev. Lett.* **54**, p. 154, 10.1103/PhysRevLett.54.154.

Shapir, Y. (1987). Static and dynamic properties of random-field systems, *Phys. Rev. B* **35**, p. 62, 10.1103/PhysRevB.35.62.

Shapira, Y. and Becerra, C. C. (1976). Phase boundaries near the bicritical point in MnF_2, *Phys. Lett.* **57A**, p. 483, 10.1016/0375-9601(76)90135-3.

Shapira, Y. and Becerra, C. C. (1977). Phase diagram of the isotropic antiferromagnet $RbMnF_3$: Test of scaling and renormalization-group calculations, *Phys. Rev. Lett.* **38**, p. 358, 10.1103/PhysRevLett.38.358.

Shapira, Y. and Oliveira, Jr., N. F. (1978). Phase diagrams of the isotropic antiferromagnets $RbMnF_3$ and $KNiF_3$ from ultrasonic measurements, *J. Appl. Phys.* **49**, p. 1374, 10.1063/1.324996.

Shapira, Y., Oliveira, Jr., N. F., and Foner, S. (1984). Effects of random-fields on the phase transitions and phase diagram of $Mn_{0.75}Zn_{0.25}F_2$, *Phys. Rev. B* **30**, p. 6639, 10.1103/PhysRevB.30.6639.

Shelton, L. J., Ye, F., Barber, W. C., Zhou, L., and Belanger, D. P. (2004). Simulation of irreversibilities of the random-field Ising model order parameter, *J. Magn. Magn. Mater.* **272–276**, p. 1302, 10.1016/j.jmmm.2003.12.084.

Slanič, Z. (1998). *Static critical behavior of the random field Ising model*, Ph.D. thesis, University of California, Santa Cruz.

Slanič, Z. and Belanger, D. P. (1998). The random-field specific heat critical behavior at high magnetic concentrations: $Fe_{0.93}Zn_{0.07}F_2$, *J. Magn. Magn. Mater.* **186**, p. 65, 10.1016/S0304-8853(98)00065-1.

Slanič, Z., Belanger, D. P., and Fernandez-Baca, J. A. (1998). Random-field critical scattering at high magnetic concentration in the Ising antiferromagnet $Fe_{0.93}Zn_{0.07}F_2$, *J. Magn. Magn. Mater.* **177–181**, p. 171, 10.1016/S0304-8853(97)00333-8.

Slanič, Z., Belanger, D. P., and Fernandez-Baca, J. A. (2001). Scaling properties of the critical behaivor in the dilute antiferromagnet $Fe_{0.93}Zn_{0.07}F_2$, *J. Phys.: Condensed Matter* **13**, p. 1711, 10.1088/0953-8984/13/8/308.

Sousa, L. L. L., Araújo, L. R. S., Machado, F. L. A., Montenegro, F. C., and Rezende, S. M. (2010). Specific heat of the dilute antiferromagnetic system $Fe_xZn_{1-x}F_2$, *J. Phys.: Conf. Series* **200**, p. 032069, 10.1088/1742-6596/200/3/032069.

Staats, M., Nowak, U., and Usadel, K. D. (1998). Non-exponential relaxation in dilute antiferromagnets, *Phase Tran.* **65**, p. 159, 10.1080/01411599808209286.

Stauffer, D. and Aharony, A. (1994). *Introduction to Percolation Theory* (Taylor and Francis).

Stauffer, D., Ferer, M., and Wortis, M. (1972). Universality of second-order transitions: The scale factor for the correlation length, *Phys. Rev. Lett.* **29**, p. 345, 10.1103/PhysRevLett.29.345.

Stout, J. W. and Catalano, E. (1955). Heat capacity of zinc fluoride from 11 to 300K. Thermodynamic functions of zinc fluoride. Entropy and heat capacity associated with the antiferromagnetic ordering of manganous fluoride, ferrous fluoride, cobaltous fluoride, and nickelous fluoride, *J. Chem. Phys.* **23**, p. 2013, 10.1063/1.1740657.

Stout, J. W. and Reed, S. A. (1954). The crystal structure of MnF_2, FeF_2, CoF_2, NiF_2, and ZnF_2, *J. Amer. Chem. Soc.* **76**, p. 5279, 10.1021/ja01650a005.

Sundaram, N., Jiang, Y., Anderson, I. E., Belanger, D. P., Booth, C. H., Bridges, F., Mitchell, J. F., Proffen, T., and Zheng, H. (2009). Local structure of $La_{1-x}Sr_xCoO_3$ determined from EXAFS and neutron pair distribution function studies, *Phys. Rev. Lett.* **102**, p. 026401, 10.1103/PhysRevLett. 102.026401.

Tarko, H. B. and Fisher, M. E. (1975). Theory of critical point scattering and correlations. III. The Ising model below T_C and in a field, *Phys. Rev. B* **11**, p. 1217, 10.1103/PhysRevB.11.1217.

Teaney, D. T., Freiser, M. J., and Stevenson, R. W. H. (1962). Discovery of a simple cubic antiferromagnet: Antiferromagnetic resonance in $RbMnF_3$, *Phys. Rev. Lett.* **9**, p. 212, 10.1103/PhysRevLett.9.212.

Terao, T. and Nakayama, T. (1995). Double-peak structure of the dynamical structure factor in diluted Heisenberg antiferromagnets, *Phys. Rev. B* **51**, p. 11479, 10.1103/PhysRevB.51.11479.

Theodorakis, P. E., Georgiou, I., and Fytas, N. G. (2013). Fluctuations and criticality in the random-field Ising model, *Phys. Rev. E* **87**, p. 032119, 10. 1103/PhysRevE.87.032119.

Tracy, C. A. and McCoy, B. M. (1975). Examination of the phenomenological scaling functions for critical scattering, *Phys. Rev. B* **12**, p. 368, 10.1103/ PhysRevB.12.368.

Villain, J. (1977). Two-level systems in a spin-glass model. I. General formalism and two-dimensional model, *J. Phys. C: Sol. St. Phys.* **10**, p. 4793, 10.1088/ 0022-3719/10/23/013.

Villain, J. (1984). Nonequilibrium critical exponents in the random-field Ising-model, *Phys. Rev. Lett.* **52**, p. 1543, 10.1103/PhysRevLett.52.1543.

Villain, J. (1985). Equilibrium critical properties of random field systems — new conjectures, *J. Physique* **46**, p. 1843, 10.1051/jphys:0198500460110184300.

Wang, J., Belanger, D. P., and Gaulin, B. D. (1991). The specific heat critical behavior of $CsMnBr_3$ and holmium: Two tests of chiral universality, *Phys. Rev. Lett.* **66**, p. 3195, 10.1103/PhysRevLett.66.3195.

Wang, J., Belanger, D. P., and Gaulin, B. D. (1992). Specific heat of $CsMnBr_3$ and Ho: $n = 2$, $d = 3$ chiral universaility, *J. Magn. Magn. Mater.* **117**, p. 356, 10.1016/0304-8853(92)90090-B.

Wang, J., Belanger, D. P., and Gaulin, B. D. (1994). Specific heat critical behavior of the Ising stacked triangular lattice antiferromagnet $CsCoBr_3$, *Phys. Rev. B* **49**, p. 12299, 10.1103/PhysRevB.49.12299.

Wansleben, S. and Landau, D. P. (1987). Dynamical critical exponent of the 3D Ising model, *J. Appl. Phys.* **61**, p. 3968, 10.1063/1.338572.

Wertheim, G. K. and Buchanan, D. N. E. (1967). Temperature dependence of the Fe^{57} hfs in FeF_2 below the Neel temperature, *Phys. Rev.* **161**, p. 478, 10. 1103/PhysRev.161.478.

Wong, P.-Z., von Molnar, S., and Dimon, P. (1982). Random-field effects in $Fe_{1-z}Mg_xCl_2$, *J. Appl. Phys.* **53**, p. 7954, 10.1063/1.330240.

Wong, P.-Z., Yoshizawa, H., and Shapiro, S. M. (1985). Coexistence of antiferromagnetism and spin-glass ordering in the Ising system $Fe_{0.55}Mg_{0.45}Cl_2$, *J. Appl. Phys.* **57**, p. 3462, 10.1063/1.335080.

Xiong, W. and Xu, C. (2019). Phase transition behavior in three-dimensional Gaussian distribution random-field Ising model with finite-time dynamics method, *J. Stat. Mech.* , p. 02320210.1088/1742-5468/aaf62f.

Yakubo, K., Terao, T., and Nakayama, T. (1994). Spin-wave dynamics of percolating Heisenberg antiferromagnets, *J. Phy. Soc. Jpn.* **63**, p. 3431, 10.1143/JPSJ.63.3431.

Yan, J.-Q., Zhou, J.-S., and Goodenough, J. B. (2004). Ferromagnetism in $LaCoO_3$, *Phys. Rev. B* **70**, p. 014402, 10.1103/PhysRevB.70.014402.

Ye, F., , Zhou, L., Meyers, S. A., Shelton, L. J., Belanger, D. P., Lu, L., Larochelle, S., and Greven, M. (2006). Quasicriticality of the order parameter of the three-dimensional random-field Ising antiferromagnet $Fe_{0.85}Zn_{0.15}F_2$: A synchrotron x-ray ascattering study, *Phys. Rev. B* **74**, p. 144431, 10.1103/PhysRevB.74.144431.

Ye, F. (2003). *Random-field Ising ordering above magnetic vacancy percolation*, Ph.D. thesis, University of California, Santa Cruz.

Ye, F. (2020). Private Communication.

Ye, F., Matsuda, M., Katano, S., Yoshizawa, H., Belanger, D. P., Seppälä, E. T., Fernandez-Baca, J. A., and Alava, M. J. (2004). Percolation fractal dimension in scattering line shapes of the random-field Ising model, *J. Magn. Magn. Mater.* **272–276**, p. 1298, 10.1016/j.jmmm.2003.12.081.

Ye, F., Rodriguez, Y. W., Belanger, D. P., and Fernandez-Baca, J. A. (2007). The order parameter critical exponent of the chiral phase transition in VF_2, *J. Magn. Magn. Mater.* **310**, p. 1410, 10.1016/j.jmmm.2006.10.424.

Ye, F., Zhou, L., Larochelle, S., Lu, L., Belanger, D. P., Greven, M., and Lederman, D. (2002). Order parameter criticality of the $d = 3$ random-field Ising magnet $Fe_{0.85}Zn_{0.15}F_2$, *Phys. Rev. Lett.* **89**, p. 157202, 10.1103/PhysRevLEtt.89.157202.

Yoshizawa, H., Cowley, R. A., Shirane, G., and Birgeneau, R. (1985). Neutron scattering study of the effect of a random field on the three-dimensional dilute antiferromagnet $Fe_{0.6}Zn_{0.4}F_2$, *Phys. Rev. B* **31**, p. 4548, 10.1103/PhysRevB.31.4548.

Yoshizawa, H., Cowley, R. A., Shirane, G., Birgeneau, R. J., Guggenheim, H. J., and Ikeda, H. (1982). Random-field effects in two-dimensional, and three-dimensional Ising antiferromagnets, *Phys. Rev. Lett.* **48**, p. 438, 10.1103/PhysRevLett.48.438.

Young, A. P. (1977). Lowering of dimensionality in phase-transitions with random fields, *J. Phys. C: Sol. St. Phys.* **10**, p. L257, 10.1088/0022-3719/10/9/007.

Index

A^+/A^-, 98, 155
D_f, 162
H_{eff}, 84
M_S, 16
R_s, 88
$T_F(H)$, 116
$T_{eq}(H)$, 120, 121, 124, 126
ΔM, 163
Δn, 33
α, 16, 81, 99, 104, 145, 155, 158
β, 45, 83, 104
β_c, 182
$\chi'(\omega)$, 127, 128
χ, 103, 106, 120
χ_S, 13
$\chi_S^{dis}(T)$, 147
η, 146
γ, 104, 146
γ_c, 182
κ, 52, 106, 120
μ, 163
ν, 16, 146
ϕ, 17, 64, 158
ϕ_c, 182
ϕ_{rf}, 123
ξ, 4, 23, 61
$d(\Delta n)/dT$, 24, 34, 76, 78, 166
$d = 2$ Heisenberg, 65
$d = 2$ Ising, 76, 77, 86
$d = 2$ Ising universality, 61

$d = 2$ random-exchange Ising, 99
$d = 3$ Heisenberg universality, 61, 64
$d = 3$ XY universality, 64
$d = 3$ random-field Ising, 125, 137
$d = 3$ random-field Ising model, 113
dC/dT, 43
d_l, 113, 121
h_{rf}, 119
v_G, 175
$x_p = 0.593$, 99
x_p, 119, 165, 166, 170, 173
$x_p = 0.246$, 99
x_v, 105, 119, 124, 133, 134, 161

absolute temperature, 24
ac specific heat, 81
ac-susceptibility, 126, 127, 130
activated dynamics, 127–129, 155
amplitude ratio, 15, 90, 107, 108, 120,
    145, 146, 183
anisotropic, 78
anomalous dilation, 137
anomalous spin diffusion, 173
antiferromagnetic, 62
antiferromagnetic Bragg point, 61
Antiferromagnetic scattering peaks,
    47
Archimedes principle, 59
asymptotic critical behavior, 4, 81
asymptotic critical region, 92

asymptotic Ising critical region, 78
asymptotic universal critical
   behavior, 15
asymptotic universal parameters, 24

bicritical, 62
bicritical point, 61, 62, 64
birefringence, 78, 144
birefringence of antiferromagnetic
   crystals, 34
body-centered cubic lattice, 174
body-centered tetragonal lattice, 58,
   99, 174
Bragg scattering, 19, 47, 147
Bragg scattering intensity, 139
Bragg scattering line widths, 89
breaking of symmetry, 34

canonical spin-glass, 169
canted antiferromagnet, 6
capacitance, 43, 159
capacitance measurements, 126
capacitance technique, 32, 124, 126
carbon-glass thermometers, 25, 30
chemical clustering or anticlustering,
   52
chemical homogeneity, 7
chiral, 180
chiral critical exponents, 183
chiral exponents, 181, 182
chiral order, 182, 183, 186
chiral universality, 179, 181, 184
chirality, 181
clustering or anticlustering, 60
Co_3O_4, 198
$Co_xZn_{1-x}F_2$, 120
Co-O-Co bonds, 198
CoF_2, 58
collimators, 46
concentration gradient, 48, 52, 106
concentration variations, 37
configuration average, 19
conjugate to the order parameter, 12
conventional dynamics, 108, 128
conventional exponents, 182

cooling and heating procedures,
   114
coordination number, 74
correlation length exponent, 162
correlation length for fluctuations, 16,
   23, 143
critical dynamics, 17
critical opalescence, 11
critical parameters, 15
critical point, 2
crossover, 17, 83, 118, 137, 190
crossover behavior, 18, 82, 183
crossover effects, 4, 23
crossover exponent, 64, 104, 123, 145,
   158, 182
crossover from pure to
   random-exchange, 105
crossover from random-exchange to
   random-field, 119, 139
crossover to random-field behavior,
   139
crystal expansion, 43
$CsCoBr_3$, 179, 180, 186
$CsMnBr_3$, 179–181, 185
$CsMnF_3$, 155
cubic anisotropy, 65
Curie-Weiss, 21, 118, 193, 194

de Almeida-Thouless boundary, 169
destroyed transition, 112
diamagnetic, 7
dielectric properties, 43, 124
dilatometry, 66
dilute antiferromagnet in a field, 118
Direct specific heat techniques, 26
disconnected susceptibility, 19, 146
dispersion, 173
domain dynamics, 161
domain formation, 5, 120, 161
domain structure, 136
domain wall dynamics, 162
domain wall energy, 5
domain wall surface tension, 113
dynamic critical behavior, 126
dynamics, 126

effective dimensionality, 130
effective exponents, 82, 149
effective parameters, 23
effective random field, 111, 118, 119
equilibrium boundary, 121, 124
equilibrium state, 120
evolution of the domain walls, 162
excess magnetization, 163
excitation spectra, 173
exponents, 15
extinction, 47, 83, 84, 89, 126, 146
extinction effects, 95, 137

Faraday rotation, 32, 41, 127
FC, 114, 115, 120, 121, 124, 126, 127, 132, 133, 138, 139, 142, 144, 149, 155, 161
FC fractal domain structure, 163
Fe^{57}, 85
$Fe_x Mg_{1-x} Cl_2$, 54, 61, 113
$Fe_x Mn_{1-x} F_2$, 59, 71
$Fe_x Zn_{1-x} F_2$, 49, 59, 64, 98, 106, 113, 119, 120, 123, 126, 136, 165, 170–172, 176
$Fe_{0.25} Zn_{0.75} F_2$, 169, 177
$Fe_{0.31} Zn_{0.69} F_2$, 172, 177
$Fe_{0.3} Mg_{0.7} Cl_2$, 169
$Fe_{0.46} Zn_{0.54} F_2$, 40, 108, 126, 127, 137
$Fe_{0.47} Zn_{0.53} F_2$, 163
$Fe_{0.68} Zn_{0.32} F_2$, 124, 126
$Fe_{0.6} Zn_{0.4} F_2$, 37, 105, 120, 163
$Fe_{0.75} Mn_{0.25} F_2$, 75
$Fe_{0.85} Zn_{0.15} F_2$, 136, 137
$Fe_{0.8} Zn_{0.2} F_2$, 51
$Fe_{0.93} Zn_{0.07} F_2$, 145, 147
$Fe_{0.9} Zn_{0.1} F_2$, 104, 108
$Fe_x Zn_{1-x} F_2$, 53
$FeCl_2$, 6, 58, 61, 65
FeF_2, 43, 49, 58, 64, 65, 75, 76, 78, 81, 84, 105, 106, 124, 163, 176
ferrimagnet, 6
ferromagnet, 6
ferromagnet iron, 5
ferromagnetic Bragg peak, 46
few impurity, 195
FH, 114, 115, 149

field-temperature cycling, 120
first-order transition, 61
first-order transition boundary, 64
fluctuation correlation length, 48, 52, 146
fluctuations, 143
fluorides, 49, 99
four-wire resistance measurements, 26
fractal, 161
fractal dimension, 162, 176
fractal lattice, 178
fractal magnetic lattice, 173
fractal nature of the domain interfaces, 162
Fractal percolation structures, 161
fractal structure, 152
fracton, 173
free energy, 113
freezing temperature, 116
Frustrated interactions, 4
fused quartz windows, 34

Griffiths singularities, 104
Ground state calculations, 161

Hamiltonian, 111
Harris criterion, 98, 99
heat leaks, 26
heat pulse techniques, 144
Heisenberg, 3, 18, 43, 64, 81
Heisenberg specific heat, 158
Heisenberg to Ising crossover, 43, 81
helical metals, 184
helical pitch, 186
helical spin order, 186
holmium, 179, 184–186
hyperfine effective field, 84
hyperscaling, 17
hysteresis, 113, 126, 127, 136–138, 144, 161

impurities or defects, 5
incommensurate, 6
instrumental resolution, 141, 146
insulating magnet, 4, 186
integrated intensity, 141

Ising, 3, 18
Ising cluster excitations, 173
isotropic antiferromagnet, 156, 159
itinerant electrons, 23

$K_2Co_xMg_{1-x}F_4$, 113
K_2CoF_4, 61, 83, 86, 88, 89, 93
K_2NiF_4, 61
$KCoF_3$, 158
$KNiF_3$, 43, 61, 65, 156, 159
Kosterlitz-Thouless, 19

$LaCoO_3$, 193, 194, 198
lattice contribution, 166
length scale for thermal fluctuations, 4
linear and nonlinear susceptibilities, 109
linear and nonlinear susceptibility, 41
linear optical birefringence, 24
linearly polarized light, 33
liquid helium, 7
liquid-gas, 10
local nonrandomness, 60
lock-in detection, 34
logarithmic divergence, 77
long relaxation, 126
long-range correlations, 143
long-range order, 113, 143
long-range random-field Ising order, 141
long-time dependence, 126
Lorentzian, 20, 148
Lorentzian line shape, 94–96, 107
Lorentzian-like, 120, 147
lower critical dimension, 113, 121

Mössbauer, 108
Mössbauer spectroscopy, 85, 104
Mössbauer techniques, 45
magnetic and vacancy percolation threshold concentrations, 60
magnetic dilution, 4
magnetic percolation threshold, 69, 165

magnetic percolation threshold concentration, 176
magnetic scattering from x-rays, 47
magnetic specific heat, 44, 75, 105
magnetically diluted, 83
Magnetometry, 42
magnetostriction, 45
magnon, 173
mean-field, 17, 18, 66, 118, 183, 193
mean-field approximation, 148
mean-field behavior, 86
mean-field line shape, 92
mean-field Lorentzian, 162
mean-field temperature shift, 118
mean-field tricritical, 184
metallic resistance thermometry, 25
metamagnet, 6
metastability, 126
metastable domains, 155, 161, 162
metastable state, 142
MgF_2, 78
microscopic relaxation time, 163
miniature coax cables, 31
mixed phase, 66
$Mn_xZn_{1-x}F_2$, 54, 59, 67, 113, 119, 120, 123, 165, 170, 172
$Mn_{0.35}Zn_{0.65}F_2$, 172
$Mn_{0.39}Zn_{0.61}F_2$, 69
$Mn_{0.4}Zn_{0.6}F_2$, 70
$Mn_{0.75}Zn_{0.25}F_2$, 66
$Mn_xZn_{1-x}F_2$, 53, 163
MnF_2, 58, 65, 66, 75, 78, 81
modified Lorentzian, 89, 91
monochromator, 46
Monte Carlo, 8, 132, 142, 187, 190

neutron, 12, 46
neutron scattering, 65, 100, 120, 126, 137, 141, 150, 186, 187
neutron scattering line shape, 15, 86, 92, 136, 149, 162, 182, 197
neutron spallation source, 46
neutron spin-echo, 48
NMR, 104
non-Lorentzian line shape, 87, 108, 120

noncritical components, 18
nuclear scattering peaks, 46

one-dimensional, 8
Onsager, 89
Onsager solution, 76
optical birefringence, 32, 156
optical linear birefringence, 35
order parameter, 12, 47, 83, 84, 126, 137, 142, 146, 187, 193
order-parameter, 144
order-parameter exponent, 136, 139
Ornstein-Zernike Lorentzian line shape, 87
overdamped, 137

parallel magnetic susceptibility, 44
paramagnetic, 62
percolating vacancy lattice, 170
percolation threshold, 169
percolation threshold concentration, 133, 173
perturbation theory, 113
phase diagram, 64
phonon background, 13, 77
phonon background contributions, 78
phonon contribution, 28, 77
piezomagnetic, 45, 86
piezomagnetism, 163, 164
platinum thermometers, 93
polarized neutron scattering, 183
polarized neutrons, 46
Powdered samples, 28
power law behavior, 15
precision resistance bridges, 26
Pressure-temperature phase diagrams, 10
proportionality between the magnetic contributions to the specific heat and $d(\Delta)/dT$, 35
proportionality between the specific heat and $d(\Delta n)/dT$, 106
pulsed heat specific heat, 13, 28
pure to random exchange crossover, 107

quarter-wave plate, 33
quartz modulator, 33
quenched disorder, 52
quenched random dilution, 98
quenched randomness, 19

random-exchange, 4, 49, 145
random-exchange antiferromagnets, 100
random-exchange dynamic critical behavior, 109
random-exchange exponents, 104
random-exchange Ising, 105
random-exchange neutron scattering, 106
random-exchange order-parameter, 104
random-exchange susceptibility, 109
random-exchange to random-field crossover, 149, 151
random-field, 4, 49
random-field critical exponents, 120
random-field crossover exponent, 118, 170
random-field dynamics, 126
random-field Ising, 17, 54, 118, 161
random-field Ising model, 110, 126, 127, 155, 170
random-field scattering, 151
random-field specific heat, 119
$Rb_2Co_xMg_{1-x}F_4$, 100, 113
$Rb_2Co_{0.58}Mg_{0.42}F_4$, 178
$Rb_2Co_{0.6}Mg_{0.4}F_4$, 178
$Rb_2Co_{0.85}Mg_{0.15}F_4$, 103, 115, 162
Rb_2CoF_4, 28, 61, 76, 83, 86, 89
$Rb_2Mn_xMg_{1-x}F_4$, 174
$Rb_2Mn_{0.4}Mg_{0.6}F_4$, 176
$Rb_2Mn_{0.7}Mg_{0.3}F_4$, 71
Rb_2MnF_4, 65
$RbCoF_4$, 178
$RbMn_xMg_{1-x}F_3$, 174
$RbMn_{0.34}Mg_{0.66}F_3$, 176
$RbMn_{0.598}Mg_{0.402}F_3$, 176
$RbMnF_3$, 61, 65, 156, 159
reciprocal space, 47, 83, 86
reduced temperature, 9, 23, 119

relative changes in temperature, 24
relative positions u, 58
relative temperature, 24
relaxation technique, 144
remanent magnetic moment, 162
remanent magnetization, 164
resistance bridges, 93
RKKY interaction, 186
rounding of the specific heat, 51

Sénarmont technique, 33
sample shield, 31
sample temperature drift, 31
sapphire, 25
scaling, 118, 134, 145, 182
scaling laws, 17
scaling relation, 105, 181
Scattering cross sections, 19
scattering line shape, 86, 125, 126, 134
second-order phase transition, 2
second-order transition, 62
short-range, 4
short-range magnetic correlations, 166
Silver paint, 28
simple cubic lattice, 174
site-dilute antiferromagnets, 54
$SmMnO_3$, 155
smooth domain walls, 162
specific heat, 16, 76, 118, 119, 127, 142–145, 183, 190
specific heat critical behavior, 159
specific heat critical behavior amplitudes, 106
specific heat peak, 143
spectrometer collimation, 83, 89
spin echo neutron scattering, 126
spin wave dispersions, 176
Spin wave excitations, 173
spin waves, 175
spin-flop, 64
spin-glass-like, 172
spin-glasses, 169
spontaneous moment, 45
spurious atomic scattering, 47

square lattice, 174
squared-Lorentzian, 20, 148, 162
squared-Lorentzian-like, 120
stacked triangular lattice antiferromagnet, 179
staggered field, 13
staggered magnetization, 83, 137, 139
staggered susceptibility, 48, 147
strongly anisotropic FeF_2, 4
structure factor, 19
superexchange interactions, 58, 60
surface, 193
surface magnetism, 193
surface ordering exponent, 193
susceptibility, 13, 103, 134, 146, 169
symmetric logarithmic, 120
symmetric logarithmic divergence, 76
symmetric logarithmic peak, 145
synchrotron scattering, 47, 137

temperature derivative of the optical birefringence, 28
temperature gradient, 52
temperature uniformity, 23
tetracritical point, 181
thermal anchoring, 26
thermal conductivity, 28
thermal cycling, 161
thermal diffusivity, 160
thermal expansion, 45
thermal fluctuations, 143, 144
Thermal grease, 28
thermal nuclear reactor, 46
thermistor, 26, 93
three-dimensional ($d = 3$), 3
three-dimensional Bragg scattering, 83
Triangular antiferromagnets, 179
tricritical, 62, 190, 191
tricritical behavior, 61, 187, 190
tricritical critical behavior, 190
tricritical point, 6, 66
triple-axis spectrometer, 46
twin interface, 198

twin interface surfaces, 198
twisted pairs, 31
two-dimensional $(d = 2)$, 3
two-dimensional scattering, 61, 83
two-scale universality, 18, 88, 89

uniaxial pressure, 157
universal parameters, 4
universality class, 15

vacancies, 52
vacancy percolation, 132, 155
vacancy percolation threshold, 136, 137, 161
vacancy percolation threshold concentration, 60

variation in concentration, 54, 126
VF_2, 179, 186

weakly anisotropic antiferromagnet MnF_2, 4

x-ray, 13, 136, 137
x-ray scattering, 46, 144, 146
XY, 4, 18, 179, 180, 182, 184, 186, 187, 190, 191
XY stacked triangular lattice, 186

ZFC, 114, 115, 120, 121, 124, 126, 127, 132, 138, 139, 142, 144, 149, 161
ZnF_2, 49, 58, 59, 78, 166